ECONOMIC ISSUES, PROBLEMS AND PERSPECTIVES

FEDERAL INFLUENCE ON INNOVATION

POLICY EFFECTS AND PROPOSALS

ECONOMIC ISSUES, PROBLEMS AND PERSPECTIVES

Additional books in this series can be found on Nova's website under the Series tab.

Additional e-books in this series can be found on Nova's website under the e-book tab.

ECONOMIC ISSUES, PROBLEMS AND PERSPECTIVES

FEDERAL INFLUENCE ON INNOVATION

POLICY EFFECTS AND PROPOSALS

RACHAEL A. YOUNG
EDITOR

nova publishers
New York

Library of Congress Cataloging-in-Publication Data

ISBN: 978-1-63463-627-8

Published by Nova Science Publishers, Inc. † New York

CONTENTS

PREFACE

In this book, the Congressional Budget Office (CBO) examines the effects on innovation of existing policies and systems and the possible effects of a variety of proposals for changing those policies and systems. This book also discusses the current status of the research tax credit; the role of trade secrets in innovation policy; and the patent system for specific industries.

Chapter 1 - Innovation is a central driver of economic growth in the United States. Workers become more productive when they can make use of improved equipment and processes, and consumers benefit when new goods and services become available or when existing ones become better or cheaper—although the transition can be disruptive to established firms and workers as new products and processes supersede old ones. Looking ahead, innovation will continue to be important for economic growth, in part because the supply of workers to the economy is expected to increase at a much slower rate in the future.

Innovation produces some benefits for society from which individual innovators are not able to profit, and, as a result, those innovators tend to underinvest in such activity. Policymakers endeavor to promote innovation to compensate for that underinvestment. The federal government influences innovation through two broad channels: spending and tax policies, and the legal and regulatory systems. In this report, the Congressional Budget Office (CBO) examines the effects on innovation of existing policies and systems and the possible effects of a variety of proposals for changing those policies and systems.

Chapter 2 - Technological innovation is a primary engine of long-term economic growth, and research and development (R&D) serves as the lifeblood of innovation. The federal government encourages businesses to invest more in R&D than they otherwise would in several ways, including a tax credit for increases in spending on qualified research above a base amount.

This report describes the current status of the credit, summarizes its legislative history, discusses policy issues it raises, and describes legislation to modify and extend it.

The research credit (also known as the research and experimentation (R&E) tax credit) has never been permanent. It expired at the end of 2011 and was retroactively extended by the American Taxpayer Relief Act of 2012 (P.L. 112-240) through the end of 2013. Since its enactment in mid-1981, the credit has been extended 15 times and significantly modified 5 times.

While the credit is usually assumed to be a single credit, it actually consists of four discrete credits: (1) a regular credit, (2) an alternative simplified credit (ASC), (3) a basic research credit, and (4) an energy research credit. A taxpayer may claim one of the first two and each of the other two, provided it meets the requirements for each.

In essence, the research credit attempts to boost business investment in basic and applied research by reducing the after-tax cost of undertaking qualified research above a base amount, which in theory approximates the amount a company would invest in R&D in the absence of the credit. As a result, the credit's effectiveness hinges on the sensitivity of the demand for this research to decreases in its cost. It is unclear from available studies how sensitive that demand actually is.

While most analysts and lawmakers endorse the use of tax incentives to generate increases in business R&D investment, some have some reservations about the current credit. Critics contend that it is not as effective as it could or should be because of certain problems with its design. These include a lack of permanence, uneven and inadequate incentive effects, non-refundability, and an ambiguous definition of qualified research.

The House has passed two bills (H.R. 4438, the American Research and Competitiveness Act of 2014 on May 9 and H.R. 4, the Jobs for America Act on September 18) that would simplify and permanently extend the research tax credit. Under each measure, the credit would be equal to the sum of 20% of a taxpayer's qualified research expenditures (QREs) in the current tax year above 50% of average annual QREs in the previous three tax years, 20% of its basic research payments in the current tax year above 50% of average annual basic research payments in the three previous tax years, and 20% of the amounts paid or incurred by the taxpayer in the current tax year for qualified energy research conducted by an energy research consortium. The bills' estimated 10- year revenue cost is $155.5 billion.

On April 3, the Senate Finance Committee marked up S. 2260, the Expiring Provisions Improvement, Reform, and Efficiency Act. Among other things, the bill would extend the existing research credit through the end of 2015. It would also change its design by allowing small, young companies that are unable to use some or all of any credit they could claim for the current tax year because of insufficient tax liability to apply up to $250,000 of their credit against their federal payroll tax liabilities. This option would be available only to companies that have been in business fewer than five years and whose annual gross receipts are less than $5 million.

Chapter 3 - Many businesses have developed proprietary information that provides a competitive advantage because it is not known to others. As the United States continues its shift to a knowledge- and service-based economy, the strength and competitiveness of domestic firms increasingly depends upon their know-how and intangible assets. Trade secrets are the form of intellectual property that protects this sort of confidential information.

Trade secret law protects secret, valuable business information from misappropriation by others. Subject matter ranging from marketing data to manufacturing know-how may be protected under the trade secret laws. Trade secret status is not limited to a fixed number of years, but endures so long as the information is valuable and maintained as a secret. A trade secret is misappropriated when it has been obtained through the abuse of a confidential relationship or improper means of acquisition.

A number of competing innovation policy concerns help shape the particular doctrines that comprise trade secret law. The availability of legal protection for trade secrets potentially promotes innovation, encourages firms to invest in employee development, and confirms standards of commercial ethics and morality. On the other hand, trade secret protection involves the suppression of information, which may hinder competition and the proper functioning of the marketplace. An overly robust trade secret law also could restrain employee mobility and promote investment in costly, but socially inefficient security measures.

Trade secrets are primarily a matter of state law. In 1996, Congress enacted the Economic Espionage Act (EEA), a statute that criminalizes both "economic espionage" and the "theft of trade secrets." The EEA provides for substantial fines and imprisonment penalties, as well as criminal forfeiture of property and court orders preserving the confidentiality of trade secrets. Some commentators believe that few prosecutions have occurred under the EEA since its enactment and have deemed the legislation ineffective.

Patents and trade secrets provide different intellectual property options for many new inventions. Inventors typically must choose (1) to maintain an invention as a trade secret, (2) to obtain a patent on the invention, or (3) to allow the invention to enter the public domain. As a result, federal legislation or other developments that are perceived to alter the effectiveness of the patent system may make the trade secret more or less attractive to industry.

As of the date of publication of this report, two bills have been introduced in the 113th Congress that focus upon trade secrets. H.R. 2466, the Private Right of Action Against Theft of Trade Secrets Act of 2013, would introduce a private cause of action within the EEA. S. 1770, the Future of American Innovation and Research Act of 2013, would establish a private cause of action against a person who misappropriates a trade secret while located outside the United States, provided that the misappropriation causes or is reasonably anticipated to cause an injury within the United States or to a U.S. person. As of the date of publication of this report, neither bill had been enacted.

Chapter 4 - Congressional interest in the patent system has been demonstrated by the enactment of the Leahy-Smith America Invents Act (AIA) in the 112th Congress. Most of the provisions of the AIA apply to any type of patented invention, whether it consists of a chemical compound, mechanical device, electrical circuit, or other technology. However, other AIA provisions are specific to particular types of inventions, including business methods, tax strategies, and human organisms. The AIA reflects the principle that, for the most part, patentable inventions are generally subject to the same statutory provisions. However, a number of exceptions exist to this concept of technological neutrality.

This blended architecture has for many years prompted inquiry into whether the patent system operates best as a uniform system that applies neutrally to all inventions, or whether it could or should be tailored to meet the specific needs of different industries. Technologies and industrial sectors arguably differ in ways salient to the patent system. Among these distinctions are the costs and risks of research and development, the availability of trade secret protection as an effective alternative to patenting, the number of patents that cover a particular product, and the patterns of patent acquisition and enforcement of firms within that sector. The patent system involves a number of parameters that could potentially be adjusted to meet the needs of individual sectors, including the speed with which applications are reviewed, the scope of exclusive rights afforded by a patent, and the term of the patent.

While some observers suggest the desirability of sector-specific patent principles, others believe them to be infeasible and unwise. They observe that legislative efforts to define particular industries may prove difficult, that attorneys may sometimes be able to draft patents artfully so as to fall within a favored category, and that U.S. industry is dynamic and resistive to a static statutory definition. In addition, U.S. membership within the World Trade Organization (WTO) may limit the ability to tailor the patent system to account for different industries and inventions, to the extent that compliance with WTO standards is desired. The WTO-administered Agreement on Trade-Related Aspects of Intellectual Property, or TRIPS Agreement, in part requires WTO member states to make patent rights available without discrimination as to the field of technology. The TRIPS Agreement admits some exceptions exist to this principle of technological neutrality, however.

Should Congress believe current circumstances to be appropriate, then no action need be taken. To the degree WTO compliance is desired, Congress could also legislate along the lines permitted by the TRIPS Agreement. Notably, although the TRIPS Agreement generally disallows discrimination with respect to technological fields, it permits distinctions on other grounds. Congress could also make use of regulatory exclusivities and other complementary intellectual property rights that the TRIPS Agreement regulates less heavily.

In: Federal Influence on Innovation
Editor: Rachael A. Young

ISBN: 978-1-63463-627-8

Chapter 1

FEDERAL POLICIES AND INNOVATION*

Nathan Musick and Philip Webre

NOTE

Numbers in the text and figures may not add up to totals because of rounding.

SUMMARY

Innovation is a central driver of economic growth in the United States. Workers become more productive when they can make use of improved equipment and processes, and consumers benefit when new goods and services become available or when existing ones become better or cheaper—although the transition can be disruptive to established firms and workers as new products and processes supersede old ones. Looking ahead, innovation will continue to be important for economic growth, in part because the supply of workers to the economy is expected to increase at a much slower rate in the future.

Innovation produces some benefits for society from which individual innovators are not able to profit, and, as a result, those innovators tend to underinvest in such activity. Policymakers endeavor to promote innovation to compensate for that underinvestment.

* This is an edited, reformatted and augmented version of a report issued by the Congressional Budget Office, November 2014.

The federal government influences innovation through two broad channels: spending and tax policies, and the legal and regulatory systems. In this report, the Congressional Budget Office (CBO) examines the effects on innovation of existing policies and systems and the possible effects of a variety of proposals for changing those policies and systems (see Summary Table 1).

HOW MIGHT CHANGES IN FEDERAL SPENDING AND TAX POLICIES PROMOTE INNOVATION?

Policymakers have a number of options for expanding the federal government's contribution to innovation. CBO examined several:

- Increase funding for federal programs that support research and development (R&D);
- Increase federal spending on education (both efforts to raise the general educational level of the workforce and efforts to support education that focuses on science, technology, engineering, and mathematics, or "STEM" fields);
- Implement tax policies that provide more incentives for private spending for R&D or other sources of innovation; and
- Increase loans or loan guarantees for firms that bring innovative technologies to market.

Using more federal resources to spur innovation, however, would entail redirecting money from other federal programs, raising taxes, increasing budget deficits, or some combination thereof.

Furthermore, the federal government is not the central actor in many areas crucial to innovation. Private firms and state and local governments spend significantly more than the federal government does for many areas of R&D and for education, respectively.

Research and Development

Economic studies have shown that federal support for R&D—particularly early-stage research—has long been very important in promoting innovation. Federal spending for R&D reached $132 billion in 2013, more than doubling its 1962 value after adjusting for inflation. As a share of gross domestic product, federal spending for R&D declined by roughly half over that period (although spending for early-stage research increased slightly). The decline in the federal share of total spending for R&D during that period is largely attributable to the expansion of private R&D and to the contraction of federal R&D associated with the end of the Cold War and the space race. Increases in federal R&D spending would be expected to boost innovation by increasing total spending for R&D, although the prospects for federal support for new products that are closer to commercialization are at best mixed. Devoting additional resources to efforts to transfer government technology to the private sector would help private innovators better utilize the specialized equipment and expertise available at federal laboratories.

Summary Table 1. Potential Federal Policy Changes to Promote Innovation That Are Considered in This Report

Policy Area and Potential Policy Change	What That Change Would Entail
Using Federal Resources	
Research and Development (R&D)	
Increase federal spending for R&D ...	... by spending more on nondefense R&D and more on defense R&D that is relevant to science and technology.
Provide more support to the private sector for the commercialization of new technologies...	... by allocating more nondefense R&D funding to commercial technology development.
Use the federal government's purchasing power to promote technology development ...	... by basing procurement decisions on the innovativeness and. commercial potential of a product's or service's underlying technology.
Education	
Increase spending on academic R&D ...	... by providing more funding for research conducted at universities.
Provide financial incentives for students to specialize in science, technology, engineering, and mathematics (STEM) fields ...	... by funding more university research grants (which provide jobs for STEM graduates) and by providing subsidized grants and loans to STEM students.

Summary Table 1. (Continued)

Policy Area and Potential Policy Change	What That Change Would Entail
Make STEM instruction more effective ...	... by increasing funding for programs that provide training to teachers of mathematics and science and that develop ways to improve the teaching of STEM .
Improve educational attainment generally ...	… by spending more to support elementary, middle, and high schools; by increasing Pell grants, lowering interest rates for postsecondary student loans, or both; or by improving school performance in other ways.
Tax Policy	
Reinstate and enhance the R&D tax credit ...	... by making it permanent, simplifying it, decreasing the potential for so-called windfalls, or making it more relevant to the current state of technology.
Increase tax incentives for domestic manufacturing ...	... by reducing the after-tax cost of investment and other activities related to innovation.
Loan and Loan Guarantee Programs	
Increase federal credit for young, innovative firms ...	... by promoting loans and loan guarantees to firms that commercialize new technologies with social benefits that are not fully captured by the market.
Legal and Regulatory Environment	
Immigration Policy	
Increase the supply of foreign-born workers in the United States who have STEM training ...	... by issuing more H-1B visas (for work on a short-term basis) and immigrant visas.
Allow more foreign-born students specializing in STEM fields to remain in the United States after they graduate from a college or university ...	... by offering them a nonimmigrant or immigrant visa.
The Patent System	
Increase the resources available to the U.S. Patent and Trademark Office (USPTO) and to federal courts ...	... by making funding for the USPTO independent of Congressional appropriations, using general revenues to supplement user-based funding for the USPTO, or providing more funding to the federal courts that adjudicate patent lawsuits.

Policy Area and Potential Policy Change	What That Change Would Entail
Reduce low-quality patents …	... by changing the ways in which the USPTO administers patent law—making patent examination, and the requirements for patent applications and renewals, more demanding.
Reduce the costs defendants incur as a result of patent litigation and, in particular, the frequency of nuisance lawsuits ...	... by changing patent law—reducing the potential payoff from engaging in patent litigation, making it harder for plaintiffs to prevail in patent infringement lawsuits, and limiting the scenarios in which patent infringement can be asserted.
Regulatory Policies and Tools	
Modify regulatory goals to favor more innovation ...	... by placing greater weight on innovation than federal regulations do today relative to other goals (such as ensuring public safety).
Adopt flexible regulatory tools ...	... by drawing on price signals and market forces to reduce the cost of regulation to innovators and by revisiting existing regulations.
Restrict state and local laws and regulations that impede innovation ...	... by forestalling regulations (for example, on product liability) that could impede innovation.

Source: Congressional Budget Office.

Note: The inclusion or exclusion of any particular option does not imply an endorsement or lack thereof by CBO, and the report makes no recommendations.

Education

A more educated workforce could spur innovation in the economy in two ways: by developing more innovative ideas and by implementing those ideas more readily. In 2013, federal spending for education, not counting student loans, totaled $126 billion, amounting to roughly 15 percent of total spending on education nationally. The effect on innovation of increased federal spending for education generally is uncertain. Over the past 40 years, per-student spending (from all sources) at the elementary and secondary levels has more than doubled in real (inflation-adjusted) terms, but student achievement has remained flat. Additional spending for STEM education might contribute to innovation. However, federal spending on STEM education programs, totaling roughly $3 billion, currently constitutes a small fraction of federal spending for education overall and is spread across many agencies and grade levels. The most significant federal contribution to the education of new scientists comes from the federal spending for university R&D, which often pays for the training of graduate students and newly minted Ph.D.s working in the laboratories of established scientists.

Tax Policy

The tax code can provide financial incentives to individuals and businesses to pursue innovation through the R&D tax credit and other tax preferences. In 2012, the R&D tax credit resulted in tax expenditures totaling $6 billion. However, the R&D credit expired in December 2013, raising the question of its extension or modification. Some analysts argue that the R&D credit stimulates some additional private R&D, whereas others contend that much, if not most, of the forgone revenues go to firms for performing R&D that they intended to perform anyway, regardless of the credit. Increasing federal support for manufacturing through increased tax deductions for investment in plant and equipment would be a very indirect way to increase innovation because only a modest share of such investment is made by manufacturing firms and a large share of manufacturing is not very R&D-intensive.

Loan and Loan Guarantee Programs

The federal government can also attempt to promote innovation by providing more loans or loan guarantees to private firms that commercialize new technologies with social benefits that are not fully reflected in the market, such as some sorts of renewable energy. Because private investors already devote substantial resources to commercializing innovative products and services, increasing financial support for such federal credit programs may be difficult without funding projects that could obtain private funding anyway or funding projects whose social costs outweigh their benefits.

How Might Changes in the Federal Legal and Regulatory Environment Promote Innovation?

To further encourage innovative activity, policymakers could also make changes to immigration policies, the patent system, and the regulatory regime. Modifying laws and regulatory policies would involve less of a commitment of federal resources than increasing federal spending.

Immigration Policy

Foreign-born workers—and particularly highly skilled immigrants—contribute significantly to innovation in the United States, partially because many are highly educated and disproportionately employed at high-tech firms, universities, and other institutions that foster innovation. However, U.S. immigration policy is more oriented toward family reunification than admitting workers who would be likely to contribute to innovation. Policymakers could modify immigration law to increase the number of highly skilled noncitizens that are allowed to enter and work in the United States and so promote innovation. Analysts disagree about whether increasing the immigration of such highly skilled workers would negatively affect the employment and wages of highly skilled native-born workers.

The Patent System

The patent system promotes innovation by helping inventors recoup the costs of their efforts in exchange for making their inventions public. The Congress is currently faced with calls to address a number of perceived shortcomings of the patent system, including the recent proliferation of supposedly low-quality patents, pronounced delays in processing patent applications, and the cost of infringement litigation (in particular, the frequency of "nuisance" lawsuits). Because modifications to the patent system could have both positive and negative effects on the incentives to innovate, and because research on important issues of concern today remains limited, estimating the net impact of specific proposals for patent reform is generally difficult.

Regulatory Policies and Tools

Policymakers could alter how federal regulations affect the pace of innovation in several ways. First, they could change the emphasis on innovation when there are tradeoffs between innovation and other federal goals, such as public safety. Separately, policymakers could rely more on regulatory tools that draw on prices and market forces to reduce some of the costs resulting from regulation. Finally, federal policymakers could address the ways in which liability laws and other regulations promulgated by state and local governments affect the balance of innovation and other policy goals.

SECTION 1. EFFECTS OF INNOVATION ON THE ECONOMY

Innovation is a key determinant of long-term economic growth: The development of new products and processes makes businesses more productive, and consumers benefit as new goods and services become available or as existing ones become more affordable. In addition to benefiting firms (in the form of higher profits) and consumers (in the form of lower prices for goods and services), innovation often makes a contribution to society's welfare that is not easy to measure economically, improving the quality of life in ways that are not reflected by increases in individuals' income. However, because innovation can also be disruptive to established firms and workers, the gains from such activity may be unevenly distributed.

Innovation and Economic Growth

Innovation can take many forms, and innovators are found throughout the economy. Broadly speaking, an innovation is a new or significantly improved product or process—for example, a breakthrough device, such as the integrated circuits that are at the heart of modern computing, or a better way of operating a business, such as a more advanced inventory management system. Both the quality of an innovation and the extent to which it is used determine the impact it will have on the economy.

Although the technological advances that have occurred in the manufacture of computer and communications equipment are prominent examples of innovative activity today, other sectors of the economy also generate important innovations. In retailing, for example, new methods of marketing and distributing products—both over the Internet and through traditional retail stores—have resulted in lower prices (and in the case of the Internet, added convenience) for consumers. Some of the benefits of innovation may not be readily measured in economic terms. For example, innovative pharmaceuticals and improved medical equipment and procedures continue to improve the quality of life and to extend life expectancy.

Because the effects of innovation on the economy can be difficult to measure, economists typically use the growth in total factor productivity (TFP) as a proxy. Growth in TFP is defined as the growth of real output that is not explained by increases in the amount of labor and capital—typically physical structures and equipment used in production, along with intangible capital such as computer software and research and development (R&D).

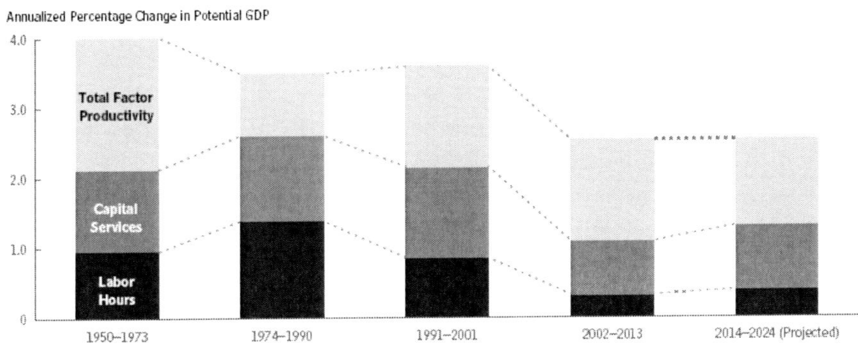

Source: Congressional Budget Office.

Notes: Potential GDP is the maximum sustainable output of the economy. This figure depicts growth in the nonfarm business sector (which excludes the economic activities of general government, private households, nonprofit organizations serving individuals, and farms).

Total factor productivity is the growth in output that remains after removing the contributions from growth in labor hours and capital services. Capital services are a measure of the flow of services available for production from the real (inflation-adjusted) stock of capital (typically, the physical structures and equipment used in production, along with intangible capital, such as computer software and knowledge gained from spending for research and development).

GDP = gross domestic product.

Figure 1-1. Contributions to the Growth of Potential GDP, 1950 to 2024.

For example, innovation can result in improvements in the quality of capital that are not reflected in the measured amount of capital used. Innovation can also allow production or distribution to be reorganized so that it becomes less expensive to supply goods and services to consumers using the same amounts of capital and labor. The increase in economic output associated with such improvements will show up as an increase in TFP and growth in the economy.[1]

The contribution of TFP to economic growth has varied over the years, as have the contributions of other factors (see Figure 1-1). From 1950 to 1973, the growth in TFP added nearly 2 percentage points each year to the growth of potential gross domestic product (potential GDP is the maximum sustainable level of output). The contribution of TFP to potential GDP was only about half as large between 1974 and 1990, although an increase in the contribution of labor hours (the number of hours worked) partially offset that decline. Over roughly the past 20 years, the contribution of TFP to potential GDP has risen to about 11/2 percentage points each year; the contributions from labor hours and capital declined over the 2002–2013 period. The Congressional Budget Office projects that the rate of growth in TFP will be slightly smaller between 2014 and 2024 than it was in the preceding two decades. The growth in the supply of workers in the economy is expected to be about as slow during the next 10 years as it was in the past decade, owing both to the retirement of baby boomers and to a relatively stable labor force participation rate among working-age women after sharp increases in some previous decades. For that reason, innovation that makes those workers more productive—for instance, by improving the equipment they use and by enabling their work to be organized more efficiently—will continue to be important.

The Impact of Innovation on Firms and Workers

Although innovation leads to economic growth, it can also lead to changes that adversely affect established firms and workers. For example, in the travel industry, the wide availability of the Internet brought with it the ability for consumers to search for information, to compare prices, and to book travel and hotel accommodations quickly, thereby making many travel agencies redundant. Many bookstores, record shops, and video rental services eventually met a similar fate. Even large firms that have become leaders in an existing technology may be unable to transfer their success to a new one such as occurred in the film manufacturing and processing industry after digital photography replaced celluloid film.

Innovation can also affect the labor force and, in particular, the distribution of income among workers in different sectors or with different skill sets. Workers in firms that go out of business as a result of a new technology have to find new jobs, which may pay less than they earned previously. In contrast, the availability of new technologies may increase the demand for workers who have the skills to use them. As a result, employment and income may rise for individuals whose training or ability enables them to make use of innovations but fall—at least as a share of the workforce and of total worker income, if not in an absolute sense—for those individuals who are less well positioned to do so.

The spread of information technology throughout the economy during the past few decades illustrates the ways in which innovation can shift the distribution of employment and income among workers. New information technologies have complemented nonroutine cognitive work (such as analysis, evaluation, and decisionmaking) and have increased the demand for the highly skilled, more educated workers who perform those tasks. As a result, both the share of employment in that kind of work and the real (inflation-adjusted) median wage earned by the most educated workers have risen over the past few decades. Beginning in the 1990s, the use of information technology not only made highly skilled workers more productive, it also increasingly substituted for workers who perform routine jobs requiring average skills (such as typing and filing). As a result, information technology led to a decline in the share of employment in occupations in which workers perform more routine cognitive tasks. In real terms, the median wage of workers with less than a bachelor's degree has displayed little or no growth.[2] Technology-driven developments in labor markets will almost certainly continue as innovation allows additional tasks—such as those required in many stages of manufacturing—to be automated.

SECTION 2. USING ADDITIONAL FEDERAL RESOURCES TO SPUR INNOVATION

The federal government promotes innovation directly by funding research and development and education, and indirectly by encouraging private investment in R&D and other innovative activity through tax preferences and loans and loan guarantees. However, federal agencies are no longer making the largest investments in some fields. The federal share of total R&D in the United States has decreased over the past several decades—both because of the expansion of private R&D and because of the diminution of federal R&D associated with the end of the Cold War and the space race. And, although the federal government provides a substantial portion of funding for academic R&D, which is used to train graduate and postdoctoral students specializing in science and related fields, most funding for education is provided by state and local governments.

Using more federal resources to spur innovation would mean either diverting money from other federal programs or activities—which would entail compromising on or reassessing national priorities—raising taxes, increasing deficits, or some combination thereof.

Funding for Research and Development

The process of R&D contributes to economic productivity by generating knowledge that leads to the development of new goods and services or to improvements in existing goods and services. (The performance of R&D also produces trained scientists, which is discussed later in this report.) For example, modern computers grew out of research into integrated circuits, and many of today's innovative services and business processes, such as Internet retailing and advanced inventory management systems, rely on advanced computer technology. The R&D process encompasses three stages:

- Development or improvement of products and services—which accounted for almost two-thirds of the R&D expenditures in the economy in 2012;
- Applied research—research designed to link scientific knowledge to some practical purpose, ultimately leading to the development of a marketable product or service—which accounted for a fifth of the national R&D; and
- Basic research—scientific inquiry that has no clear-cut commercial application but is nonetheless valued for the knowledge it produces and its potential for leading to future discoveries—which accounted for a sixth of such expenditures.

Economic analysis suggests that the benefits from R&D would justify higher spending on R&D than would be undertaken by the private sector alone. As mentioned above, the benefits that flow from R&D typically extend beyond the profits gained by the firm or person producing the knowledge, reducing the incentive to invest in the production of such knowledge relative to the amount of social benefit. That difference in incentives, particularly in basic research, has long provided a rationale for federal involvement. In addition, federal agencies—most notably the Department of Defense (DoD) and the National Aeronautics and Space Administration (NASA), but others as well—often need to develop specialized technology and equipment in order to fulfill their assigned missions.

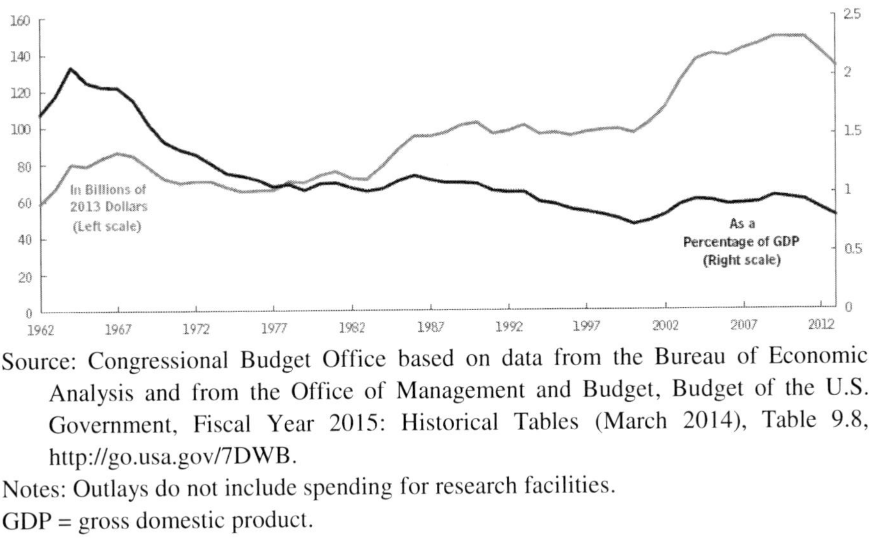

Source: Congressional Budget Office based on data from the Bureau of Economic Analysis and from the Office of Management and Budget, Budget of the U.S. Government, Fiscal Year 2015: Historical Tables (March 2014), Table 9.8, http://go.usa.gov/7DWB.

Notes: Outlays do not include spending for research facilities.

GDP = gross domestic product.

Figure 2-1. Federal Outlays for the Conduct of Research and Development.

Federal Spending for R&D

Federal outlays for R&D more than doubled between 1962 and 2013 in real terms, driven mostly by an increase in spending for R&D related to defense during the defense buildup of the early 1980s and by an increase in spending for health research from 1998 to 2004.[1] In 1962, federal outlays for R&D totaled about $59 billion (in 2013 dollars). By 2013, federal agencies were spending $132 billion for R&D (see Figure 2-1), averaging a 1.6 percent real annual rate of growth between 1962 and 2013. However, growth has been intermittent. For example, since 2009, real federal R&D spending has declined by about 10 percent. When measured against the growth of gross domestic product, federal spending for R&D has generally declined since the 1960s (with the notable exceptions of the two periods discussed above).

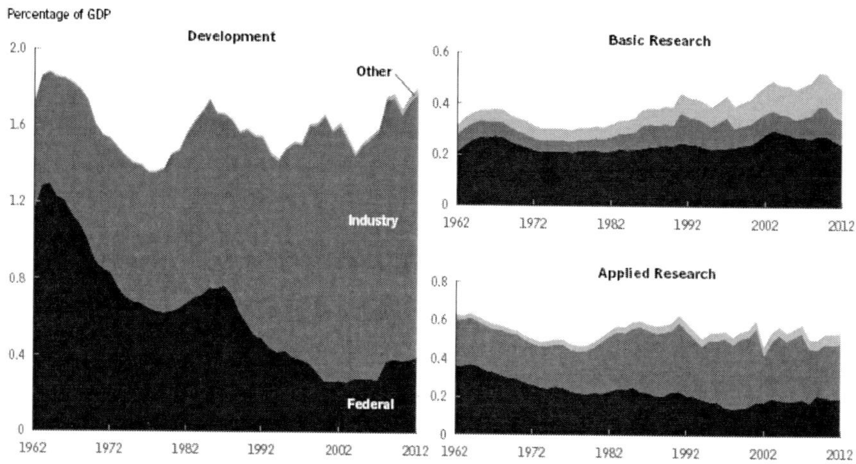

Source: Congressional Budget Office based on data from the National Science Foundation, "National Patterns of R&D Resources: 2011–12 Data Update" (December 2013), http://go.usa.gov/EWBH.

Notes: The category "Other" consists of support from universities, colleges, nonprofit organizations, and state and local governments but excludes funding from the federal government or industry.

Basic research is scientific inquiry that has no clear-cut commercial application but is nonetheless valued for the knowledge it produces and its potential for leading to future discoveries. The objective of applied research is to link scientific knowledge to some practical purpose, ultimately leading to the development of a marketable product or service.

Estimates from the National Science Foundation for national expenditures for research and development (R&D) are based on surveys of performers of R&D and thus differ slightly from estimates of outlays that come from federal agencies. Data for 2012 are preliminary.

GDP = gross domestic product.

Figure 2-2. Expenditures for Research and Development as a Share of GDP, by Source of Funding.

Although federal R&D spending has been declining as a share of the economy overall, R&D expenditures from all sources reached 2.8 percent of

GDP over the past several years, a level that had not been reached since the early 1960s.[2] Since 1962, the share of GDP devoted to basic research expenditures has increased by half, whereas the share of applied research expenditures has declined slightly. The share of development expenditures has fluctuated around its average (see Figure 2-2).

Looking at the different stages of R&D and sources of spending, expenditures by private industry across all stages have grown in the past 50 years. The largest single shift was the increase in industrial development expenditures—up by 150 percent relative to the size of the economy—that almost completely offset the decrease in federal expenditures for development (which were mainly for activities related to defense and space exploration).

Federal agencies also decreased their expenditures on applied research as a share of GDP by 45 percent, and an increase in spending by industry offset part of that decline. Federal agencies did increase their expenditures on basic research slightly relative to the size of the economy, at the same time that other nonindustrial institutions such as universities, nonprofits, and state and local governments increased their expenditure share by almost twice as much. In the early 1960s, federal expenditures accounted for roughly two-thirds of all spending devoted to R&D. Industry and all other sources combined (universities and colleges, nonprofits, and state and local governments) accounted for roughly one-third of total R&D expenditures. Since 2000, the shares have been reversed: Industry has accounted for roughly two-thirds of national R&D expenditures; the federal government has accounted for between 25 percent and 30 percent; and other sources, collectively, have accounted for 7 percent.

Trends in Federal Spending for R&D by Subject Area

Viewing federal spending for R&D by budget function shows that two areas accounted for roughly 80 percent of federal R&D spending in 2013. Defense-related R&D accounted for 54 percent of all federal R&D outlays, and health R&D accounted for an additional 24 percent. The remaining 22 percent was spread among all other categories of nondefense R&D, such as agriculture, energy, general science, and space exploration. Defense and health have grown the most over the past few decades (see Figure 2-3).

The defense category primarily supports R&D undertaken by DoD to develop weapon systems and to support science and technology.[3]

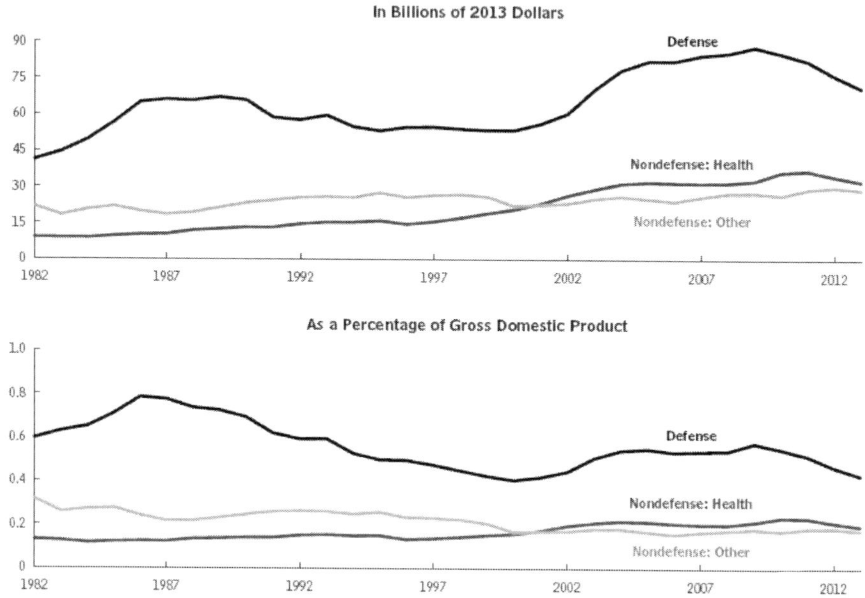

Source: Congressional Budget Office based on data from the Bureau of Economic Analysis and from the Office of Management and Budget, Budget of the U.S. Government, Fiscal Year 2015: Historical Tables (March 2014), Table 9.8, http://go.usa.gov/7DWB.

Figure 2-3. Federal Outlays for Research and Development, by Subject Area.

In 2013, DoD received $66 billion for R&D, of which 82 percent was for weapons development. Over the past decade, funding for weapons development by DoD has risen dramatically and then fallen, growing from $59 billion (in 2013 dollars) in 2003 to a peak of $73 billion in 2008 and then declining to $54 billion in 2013.[4] By contrast, real funding for DoD's science and technology programs has generally varied by less than 10 percent from its 2004–2013 average of $14 billion. However, in 2013, DoD science and technology R&D fell 18 percent below that average (other federal R&D programs declined as well).

Most growth in spending for nondefense R&D has been in the area of health. Health-related R&D has grown notably in real terms over the past few decades, whereas the growth in spending for R&D in other areas has been limited. Support for health R&D received a very large boost from 1998 to 2004 when the Congress passed legislation to double appropriations for the National Institutes of Health (NIH). Following that surge in funding, NIH's appropriations began to decline in real terms. That decline resulted in substantial disruption among biomedical research institutions that had committed to higher growth in spending for facilities and the training of young scientists, which they then found difficult to maintain.

The different areas of R&D funded by the federal government can vary in their effects on innovation and productivity growth. The two categories of nondefense R&D—health and other—and the small portion of defense R&D that is directed largely at basic and applied research probably have a positive effect on private-sector productivity.[5] Defense R&D devoted to weapons development is often thought to have less applicability to innovation in commercial or civilian products (notwithstanding some prominent technological advances that have resulted from defense R&D and subsequent procurement of weapons arising from the R&D process).[6]

R&D for national defense and health also produces public benefits that are not fully captured in the accounting of GDP. Economists have long estimated that federal health R&D has had a substantial impact on life expectancy, public health, and society's well-being.[7] However, such social benefits are often only indirectly reflected in the output of workers and calculations of national income and economic growth.

Although federal and industrial spending on health-related R&D continues to yield new medical discoveries and new drugs, the average cost for each new drug or medical discovery is rising. For instance, each extra billion dollars spent on pharmaceutical R&D now results in fewer new drugs than was the case in years past.[8] Potential explanations for this phenomenon range widely—from the likelihood that researchers have exhausted the better understood molecular targets to the effects of corporate mergers among pharmaceutical companies. In response to the decreasing number of new drugs and treatments that result from health-related R&D, NIH has an initiative specifically devoted to addressing that issue.

Policies to Increase Federal Support for R&D

The decline in the federal share of R&D relative both to private efforts and to the size of the economy has led some analysts to suggest that the federal government could increase innovation by increasing its overall R&D budget. Some economic analysis suggests that the benefits from R&D would justify much higher spending on R&D and, in particular, much higher spending on basic research.[9]

However, in an effort to limit overall federal spending, policymakers have placed caps on most discretionary funding, a category of the federal budget that includes appropriations for R&D. Discretionary programs—that is, programs funded though annual appropriation acts— run the gamut from agriculture to youth opportunity grants and include most federal programs other than benefit programs (such as Social Security and Medicare) and interest payments on the national debt. Given the caps on most funding for those programs through 2021 that are in current law, the Congressional Budget Office projects that discretionary spending will shrink markedly as a share of GDP in coming years (see Figure 2-4).

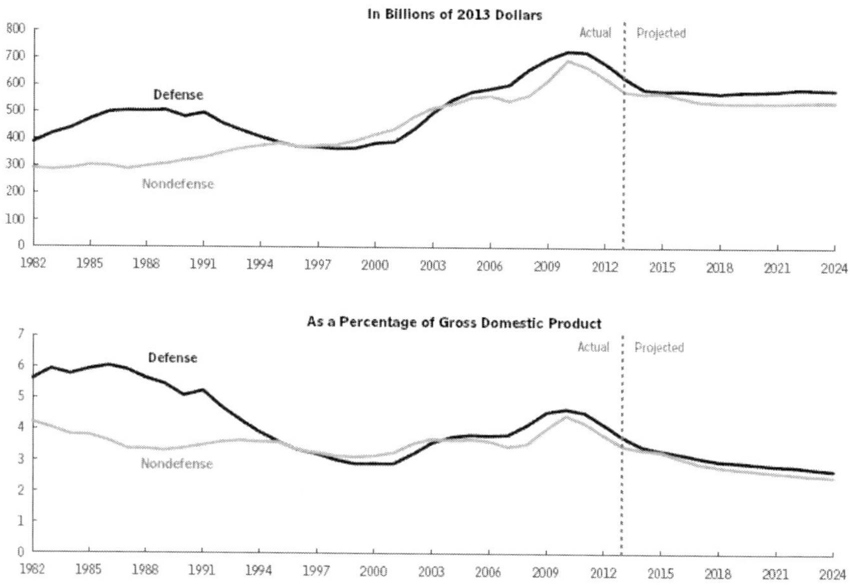

Source: Congressional Budget Office based on data from the Bureau of Economic Analysis and from the Office of Management and Budget, Budget of the U.S. Government, Fiscal Year 2015: Historical Tables (March 2014), Table 8.7, http://go.usa.gov/7DWB.

Figure 2-4. Actual and Projected Federal Outlays for Discretionary Programs.

If policymakers choose to maintain spending for R&D at its historical share of total discretionary spending, then federal R&D would be expected to shrink significantly relative to the size of the economy. Thus, given the current budgetary caps, policymakers would have to reduce the share of discretionary funding going to other kinds of activities if they decided to increase R&D

funding. In the past, federal R&D has accounted for a fairly stable share of discretionary spending. Since 1985, nondefense R&D has typically accounted for between 9 percent and 11 percent of nondefense discretionary outlays, and defense R&D has typically ranged between 12 percent and 15 percent of defense discretionary outlays. CBO projects that in the 2020s DoD will not have the budgetary resources to fund its current plans, so that sustaining or increasing R&D would require disproportionate cuts to other areas.[10] Similarly, about half of the nondefense portion of discretionary spending is for investment of some kind, whether R&D, education (discussed in the next section of the report), or infrastructure.[11] Increasing funding for R&D would put greater budgetary pressure on other categories of federal investment or on nondefense discretionary activities that do not constitute investment.

Policies to Redirect Federal Resources for R&D

Some advocates of a greater federal contribution to innovation propose that the government spend more not on R&D but on programs that would translate the results of research into commercial products. Looking back to earlier generations of successful government technology programs, some analysts further suggest that the federal government should use its purchasing power to increase the demand for products that might have substantial commercial spinoffs.

Federal Programs to Commercialize Technology

Over the past 30 or more years, the Congress has enacted various laws and funded programs designed to facilitate cooperative R&D between and among government, industry, and academia with the intent of increasing commercial innovation. The goals of the programs are varied: Some target new energy technologies, others are directed at small businesses, and others are geared to specific industries or technologies. Similarly, the programs use a variety of vehicles to accomplish those goals. Overall, federal programs to commercialize technology through product development have had a mixed record. However, programs that focus on developing technology—while leaving industrial partners to handle commercial aspects—generally seem to have enjoyed more success in the past.

Although some federal programs that were designed to commercialize technology have had a measure of success, others have proved to be economically redundant at best and commercial failures at worst. In some areas, such as agriculture and health, government agencies have long assisted industry with the introduction of innovative products, such as new plant varieties, the annual flu vaccine, or the development of drugs for AIDS. Other fields have proved more problematic. For example, in coal-powered electricity and commercial nuclear power, the Department of Energy (DOE) has long sought to introduce new energy production technologies through expensive technology demonstration plants that often have failed to deliver commercially useful products or to attract much private investment.[12]

Efforts to transfer technology out of federal laboratories to private-sector partners seem to have borne some fruit and are more numerous than some observers may realize. Many innovative commercial products incorporate technology developed at federal laboratories or using federal research funding. One survey of award-winning innovative products showed that, over the past 20 years, technologies developed at federal laboratories—often in collaboration with industry—have accounted for roughly a third of such products introduced every year.[13] Nor is such transfer of technology from federal laboratories to industry limited to the award-winners: Every year roughly 7,500 formal cooperative research ventures take place that involve many of the 300 federal laboratories active in technology transfer and either industry or universities active in developing the commercial potential of technology.[14] In addition, those federal laboratories have entered into a rising number of less formal collaborative R&D agreements with outside parties, totaling 25,000 in 2011 (the most recent year for which estimates are available). Similarly, the Small Business Innovation Research and Small Business Technology Transfer programs provide research grants to roughly 6,500 small businesses per year to help with the commercialization of new technologies.[15] In all, federal agencies fund or participate in some 40,000 individual cooperative efforts to commercialize new technologies yearly, independent of federal agencies' researching or funding technology development that might have commercial potential.

At least part of the problem with large federal technology demonstration programs—as opposed to the transfer of individual inventions developed in federal laboratories— is the fact that federal agencies are not usually equipped to oversee the commercial and business aspects of new product introduction. Companies launch new products and services embodying novel technology daily. The commercial success of those products and services often depends less on the novel technologies they incorporate and more on product design and broader business factors—the timing of product introduction, market placement, advertising, cost control, logistics, and product support. In those areas, federal employees generally have no special expertise; indeed, the very mission of many federal agencies often militates against such considerations. Those factors suggest that federal agencies may be ill-suited to participate in expensive demonstration projects; but they may experience more success with smaller, peer-reviewed efforts in basic and applied research where the technological competence predominates.

Use of Federal Procurement to Promote Technology

One approach to commercializing federally sponsored technology would involve using government purchases of new technologies to reduce their cost and help them compete successfully in the open market. In the early 1960s, DoD and NASA purchased large numbers of integrated circuits. Because those government purchases were large enough for firms to achieve economies of scale and reduce per-unit production costs in that industry, those purchases paved the way for the nation's successful integrated circuit industry. In the 1980s, both the Internet and the Global Positioning System made the transition from a military-based technology infrastructure to one primarily serving private demand. Advocates of such policies argue that the federal government could try to accomplish the same thing today in several areas—most notably in the energy sector.

However, recent efforts along those lines have generally not been successful. For example, in trying to reduce carbon emissions produced by coal-fired power plants and other industrial facilities, DOE is sponsoring the construction of power plants that separate carbon dioxide from the exhaust, compress it into a fluid, and then pump it into deep underground geologic formations (a process commonly called carbon capture and storage). DOE hopes that the cost of those plants will fall enough to make them competitive with conventional coal-fired plants. At least one problem with this strategy is that DOE's program is not nearly large enough to reduce the costs of building such plants, given normal patterns of industrial cost reduction.[16]

The federal government would have to spend perhaps tens or hundreds of billions of dollars to make a substantive impact on a number of industries on the basis of procurement alone; that kind of additional discretionary spending is unlikely to be forthcoming. A spending increase of smaller size devoted specifically to procurement would be too small to be useful to spur innovation in most fields in a systematic way. Furthermore, federal procurement increasingly utilizes commercial off-theshelf technology to help reduce costs. DoD, the largest single federal purchaser of advanced technology goods, has made a policy decision to use commercial off-theshelf technology where available, including in the computer market.[17] As a result of increases in civilian demand for high-technology goods, technology is now more advanced and costs are often lower in civilian products than in comparable defense products.

SUPPORT FOR EDUCATION

A more educated workforce could spur innovation in the economy in two ways: by developing more innovative ideas and by putting those ideas into use more readily. Highly trained individuals are most likely to come up with the insights that will drive innovation, creating new products and services and making existing ones better. For example, the share of people working in corporate research laboratories who have completed graduate or even postgraduate training has increased over time. Research also shows that more educated workers have a comparative advantage at making use of new technologies. At present, governments at all levels and private sources currently spend over $1.1 trillion a year on education.

Federal Spending for Education

Federal spending for education (at the primary, secondary, and postsecondary levels) and training has risen substantially since 1982, both in real terms and as a share of the economy. In 1982, federal spending for education, excluding student loans, was $41 billion (in 2013 dollars). Under the American Recovery and Reinvestment Act of 2009, funding for the Federal Pell Grant Program and for education in general increased substantially. However, that additional spending has ended, and policymakers have tightened eligibility for Pell grants and eliminated grants for summer students; as a result, federal spending on education has fallen from its peak of $160 billion in 2010 to $126 billion in 2013 (see Figure 2-5).[18] More recently, lawmakers also made several changes to federal student financial aid programs that, on net, led to relatively small changes in federal education spending.

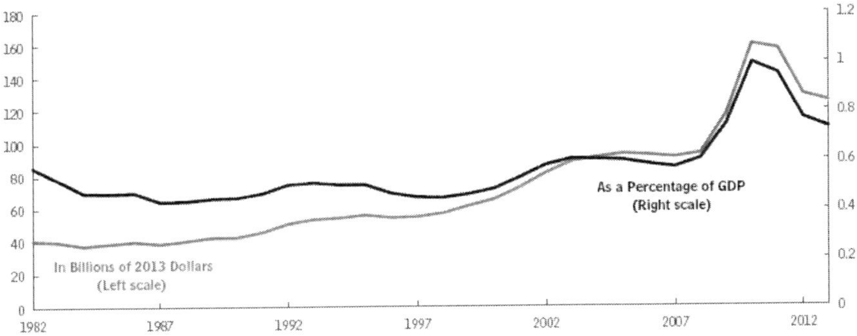

Source: Congressional Budget Office based on data from the Bureau of Economic Analysis and from the Office of Management and Budget, Budget of the U.S. Government, Fiscal Year 2015: Historical Tables (March 2014), Table 9.9, http://go.usa.gov/7DWB and Budget of the U.S. Government, Fiscal Year 2014: Supplemental Materials (April 2013), "Public Budget Database, Outlays," http://go.usa.gov/Wwy9 (XLS, 2.84 MB).

Notes: This figure does not reflect outlays for college student loans or for federal contracts or grants for university research and development.

GDP = gross domestic product.

Figure 2-5. Federal Outlays for Education and Training.

Most funding for primary and secondary education in the United States occurs at the state and local government levels. For the 2009–2010 school year, the most recent year for which data are available, revenues devoted to public primary and secondary schools totaled $606 billion (in 2013 dollars). Federal agencies provided only 13 percent of those funds ($77 billion); by contrast, 87 percent came from state and local governments and other sources (see the first chart in Figure 2-6).[19] Federal assistance at the K–12 level occurs primarily through grants for education of the disadvantaged and the disabled and for school-lunch programs.

Similarly, at the postsecondary level, most funding for education comes from sources other than federal grants. Of the $524 billion in total funding for postsecondary education during the 2009–2010 school year (in 2013 dollars), federal grants and contracts provided about 17 percent or roughly $90 billion (see the second chart in Figure 2-6). Many of the grants and contracts were for academic research, but about $32 billion was for Pell grants to help low- and lower-middle-income students attend college.

In addition to grants and contracts, the federal government issued about $100 billion in federal student loans for the 2009–2010 academic year.

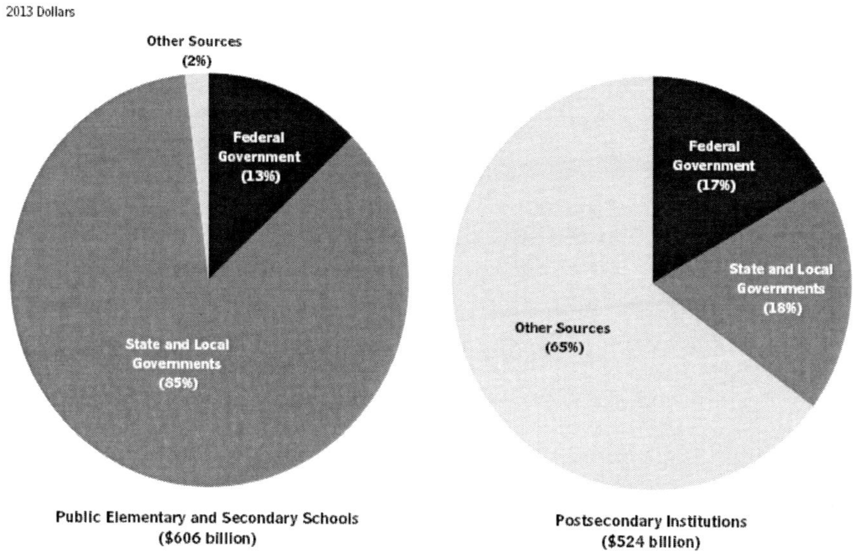

Source: Congressional Budget Office based on data from the Department of Education, National Center for Education Statistics, Digest of Education Statistics: 2012 (December 2013), http://go.usa.gov/WVkw; Office of Postsecondary Education, 2009–2010 Federal Pell Grant End-of-Year Report (last modified August 11, 2011), http://go.usa.gov/ZaGY; and National Center for Education Statistics, 2007–08 National Postsecondary Student Aid Study (April 2009), http://go.usa.gov/7rSe (PDF, 475 KB).

Notes: The 2009–2010 academic year covers the period from July 1, 2009, through June 30, 2010.

Data were not available for private elementary and secondary schools, which account for 10 percent of total enrollment in elementary and secondary schools. The numbers shown for postsecondary institutions, however, do include private schools (both nonprofit and for-profit) as well as public ones. The postsecondary institutions' revenues include support for research and development. The postsecondary institutions' "other sources" of revenues include tuition and fees (financed in part by student loans), income from assets, revenues of hospitals operated by the institutions (including amounts appropriated by governments for the hospitals), payments for services provided by the institutions (such as food services and intercollegiate athletics), and contributions from private donors.

Figure 2-6. Revenues for Educational Institutions During the 2009–2010 Academic Year, by Source.

Those loans have to be repaid by students and their families. Borrowers generally receive more favorable terms on those loans than they would receive from a private lender and, in some cases, federal loans may be their only means of paying for their postsecondary education. Yet despite those benefits to borrowers, student loans are recorded as producing budgetary savings for the government: For example, the Department of Education attributed approximately $20 billion in budgetary savings to loans made in fiscal year 2010, the year that most closely corresponds to the 2009–2010 academic year.[20] Those budgetary savings are estimated using the procedures specified by the Federal Credit Reform Act of 1990 (FCRA). If the budgetary costs or savings associated with federal student loans were estimated on a fair-value basis—an accounting method that provides a more comprehensive estimate of the cost of those loans, in CBO's view—those loans would be reported as having either costs or considerably smaller savings.[21] Consequently, a complete accounting of the federal share of financing for postsecondary education is difficult.

Federal support for adult education, which is included in Figure 2-5, consists primarily of funding for training and employment programs ($2 billion in 2013) and for veterans' education, training, and rehabilitation ($13 billion in 2013). Other spending for adult education, as when a worker takes courses at a community college to qualify for a better job, is already captured in the statistics of postsecondary education presented above.

Federal policy also supports educational investments through tax preferences. In 2014, such education-related preferences cost the Treasury $46 billion, most of which was associated with postsecondary education.[22]

Tax credits for tuition for postsecondary education alone resulted in $23 billion in forgone revenue.

Federal Spending for Education in the Fields of Science, Technology, Engineering, and Mathematics (STEM)

The federal spending that is specifically identified as supporting STEM education accounts for only a small fraction of overall federal spending for education—about $3 billion—although additional amounts are scattered among federal programs not explicitly targeting STEM education.[23] About one-third of federal STEM spending is targeted at the elementary and secondary levels and two-thirds at the postsecondary level. Federal spending for STEM education currently is channeled through more than 200 small programs.[24]

The unique federal contribution to the training of the workforce in STEM fields mainly occurs through R&D grants to universities and other research institutions; that funding is not included in the $3 billion referenced above. Graduate and postgraduate training, especially at universities that grant doctorates, includes work on faculty research projects, which is funded more often than not by the National Science Foundation (NSF), NIH, or another federal agency.[25] That work is analogous to an apprenticeship, wherein the students learn the skills necessary to engage in scientific pursuits in their chosen field. In this way, the NSF, NIH, and other agencies funding academic science contribute significantly to STEM education. Since 2000, research grants from federal agencies have paid for 61 percent of all academic R&D; consequently, federal funding underlies many of the apprenticeships that train young scientists.[26]

Policies to Increase Federal Support for Education in STEM-Related Fields

Additional targeted support for education in fields especially relevant to innovation, such as STEM, might have a direct impact on innovation, and many advocates have proposed new initiatives in those areas. However, as with R&D, most federal spending for education is funded through annual appropriation acts, so efforts to increase federal support for education in STEM-related fields would encounter the same budgetary pressures that apply to spending for R&D.

Among the specific approaches to increase federal support for STEM education are proposals to:

- Increase federal funding for academic R&D,
- Provide more scholarships and financial aid for STEM graduate students,
- Focus more STEM resources on community colleges and technical schools,
- Increase the number of mathematics and science teachers in elementary and secondary schools and improve their effectiveness through additional training, and

- Develop innovative approaches to teaching mathematics and science at elementary and secondary schools.

Additional federal funding for academic R&D may expand the size of the technically trained workforce. One recent study found that increased funding for R&D conducted by NIH led to a rise in the number of students who wanted to attend graduate school in biomedical fields. In addition, demand for such students as graduate assistants increased.[27]

An alternative to additional funding for academic R&D would be to increase funding for STEM graduate students in particular. A smaller share of people with baccalaureates in science currently receives direct federal support for graduate studies in scientific fields than was the case from the 1950s through the 1970s.[28]

Furthermore, providing additional financial support for graduate students specializing in science—either by increasing the number of grants awarded or by increasing their value—has been shown to boost the number of U.S. citizens interested in entering those fields. At the same time, some analysts argue that the U.S. educational system produces more STEM-educated workers than can find employment in those fields.[29] Roughly half of STEM degree holders at all levels and a third of people with STEM graduate degrees do not work in a STEM-related occupation.[30] Other analysts identify a wage premium paid to workers with STEM credentials, relative to wages paid for jobs with similar educational requirements, and cite it as evidence of a need for a larger STEM workforce.[31] Those wage premiums (for similar levels of educational attainment) vary substantially within STEM fields—for instance, jobs in life sciences command less of a wage premium, whereas jobs in engineering and computer fields get more.

Another alternative would be to reallocate resources within current STEM programs. At present, only 7 percent of NSF spending on STEM education focuses on community colleges and technical schools.[32] Such workers have the potential to account for half of the STEM workforce in the United States. Increasing the quality of their STEM education may increase their ability to implement innovations in the workplace. Furthermore, many community college students transfer to four-year institutions, and increasing the quality of their STEM education during the community college period is likely to result in greater academic success after the transfer.

Although some advocates of increased spending for K–12 STEM education maintain that the nation needs more teachers specializing in STEM fields, surveys of the market find that the shortages are limited to schools with high levels of student discipline problems and low levels of teacher autonomy.[33] Teachers at those schools leave more frequently than teachers at other schools. A more direct solution might be to decrease the turnover of teachers in the problem schools, for instance, by changing the financial rewards of teaching science, technology, engineering, and mathematics in those schools; however, pay and tenure policies are mainly local decisions.

Some policymakers advocate increasing the quality of the corps of STEM teachers as a way of improving the quality of STEM education. Performance by U.S. students on international tests has lagged behind the performance of students in other countries—as shown most recently by the results of the 2012 Programme for International Student Assessment.[34] Many analysts feel that the weak performance is further evidence of the need for better teachers. However, increasing the quality of the U.S. teacher corps is not straightforward: The majority of public school science teachers are already credentialed in the field, either because they majored in the field in college or received postgraduate certification.[35] Furthermore, researchers have not found evidence that the level of teachers' educational credentials is associated with gains in the educational achievement of their students. Some policymakers advocate increasing the amount of clinical experience that prospective teachers get while obtaining their education because research shows that beginning teachers are less effective than those with just a few years of experience. That evidence notwithstanding, the nation already spends much more money on education per student than other countries with higher student assessment scores.

Reform of U.S. science curricula is currently under way as part of the Administration's Race to the Top initiative, a $4 billion effort to spur innovation and improvements in K–12 education at the state and local levels. Independently, many states are already acting to improve their STEM curricula, although implementation of new educational standards may require additional time and resources.

Policies to Increase Federal Support for Education

Additional support for education could foster innovation by enhancing individuals' ability to learn and apply knowledge generally. However, it is not clear what the best approach is for adding such support. Among the potential approaches to increasing federal support for education overall are proposals to lower interest rates on student loans and to provide more generous Pell grants for lower-income students. However, federal funding for education represented 15 percent of spending for education nationwide in the 2009–2010 academic year. Consequently, even large percentage increases in federal support for education would not represent a large boost in total educational resources. In addition, over the past 40 years, spending per student at the elementary and secondary levels has more than doubled in real terms, whereas achievement among elementary and secondary students has remained flat—a result that raises questions about how effective additional spending would be in boosting achievement.

Changing the financial incentives that federal programs offer state and local educational authorities in an attempt to increase the effectiveness of federal support would be an alternative to providing more federal resources. The No Child Left Behind Act of 2001 gave states more flexibility in how they used federal funds in return for being accountable for showing progress in students' educational achievement. However, the standards used to measure that progress may have been too high, and 42 states and the District of Columbia have asked the Department of Education for waivers from the consequences of not meeting those progress standards. The Department of Education's Race to the Top program provides additional funds to states through a competitive process that also encourages states to make such reforms. To date, the evidence regarding the effectiveness of such incentives in raising students' achievement is inconclusive.

Tax Treatment of Private Investment Related to Innovation

Discussions of federal tax policy related to innovation generally focus on two major areas: using tax credits to encourage increased R&D expenditures by the private sector and using tax preferences to foster more manufacturing in the United States. Although providing a tax credit for increasing R&D expenditures has been generally correlated with increases in private R&D, many analyses find that the forgone revenues often rewarded firms for undertaking R&D that they would have performed anyway. Manufacturing accounts for a disproportionate share of private R&D, but only a small share of manufacturing is R&D-intensive, and some service-sector activity is as well, so increasing support for manufacturing generally would be less effective than increasing support for innovation directly. The sections that follow first discuss current policy in those two areas and then present options for changing those policies. (The focus on R&D and closely related investments in this section does not mean that such investments are the only private activities promoting innovation. As noted above, firms also innovate in many ways other than through R&D and closely related investments.)

Tax Credits for Increasing R&D Expenditures

Since 1981, the tax code has usually included tax credits for firms and individuals for increasing their investment in R&D.[36] However, the R&D credit has not been a permanent part of the tax code. Since it was first temporarily enacted in 1981, the credit has expired and been temporarily reauthorized multiple times (sometimes retroactively after expiring).[37] Most recently, the tax credit expired in December 2013. While in effect, the credit could be claimed for increases in spending for salaries of research personnel and supplies; it excluded expenditures for equipment and structures used for R&D as well as for fringe benefits and overhead.

In 2012, the forgone tax revenues related to the R&D tax credit were estimated to be $6 billion (in 2013 dollars).[38] Historically, the value of the R&D credit has equaled between 3 percent and 4 percent of overall spending by business for R&D.[39]

For 2013, taxpayers could calculate their credit in either of two ways:

- Using the regular credit, which was equal to 20 percent of qualified spending that exceeded a 1984– 1988 base, or
- Using an alternative, simplified credit, which was equal to 14 percent of qualified spending above 50 percent of the average of three previous years, and which was most useful for younger or merged firms or for firms for which the 1984–1988 base was otherwise irrelevant.[40]

Research-intensive industries, such as electronics and pharmaceuticals, have seen a plethora of mergers, spinoffs, and new firms such that, for the purposes of calculating the credit, the 1984–1988 base years lost relevance for many firms. Consequently, almost 60 percent of the credits for 2011, the most recent year for which data are available, were claimed for the alternative credit. Because of the incremental nature of the credit—it applied to *increases* in qualified R&D spending above a specific base amount—calculating the credit was complicated.

Despite the credit's temporary nature, restrictions, and complexity, its use expanded over the past decade. Generally, credit claims tracked the rise, decline, and subsequent resurgence of private-sector R&D during that period. Soon after its passage in 1981, economic analyses suggested that private-sector R&D was not very responsive to the credit. More recent analyses have suggested a higher level of responsiveness to the credit by firms conducting R&D.[41] One interpretation of that change is that firms had to learn how to best use the credit. However, it may be difficult to determine whether firms truly increased their R&D efforts or whether firms reclassified existing expenses as related to R&D to improve the tax treatment of such activities.[42] Other research suggests that, even if firms responded to the credit, much—or even most—of the forgone federal revenues resulted from firms' claiming credits for performing R&D that they intended to conduct anyway, regardless of the credit.[43]

Tax Preferences for Domestic Manufacturing

Manufacturing firms account for an outsized share of the nation's private-sector R&D—two-thirds of R&D compared to little more than one-fifth of gross output in 2011. Some analysts associate innovation with a robust manufacturing sector for reasons unassociated with that sector's R&D spending; for example, knowing how to build a product, or having ready access to people who do, can make it easier to improve upon the product.[44] From that perspective, encouraging domestic manufacturing would also foster innovation.

One way that federal policy encourages manufacturing is by lowering the after-tax cost of investment. Manufacturers and other firms are generally allowed to claim accelerated deductions from their taxable income each year for costs associated with the depreciation of plants and equipment located in the United States; that depreciation is defined as resulting from wear and tear, deterioration, and obsolescence. Those accelerated deductions encourage investment by enabling firms to more quickly recover the cost of their spending on physical capital. An additional first-year depreciation (known as bonus depreciation) was allowed in recent years for equipment spending in order to provide an extra incentive for investment. (The most recent provision was in effect from January 1, 2012, to December 31, 2013.) Finally, firms with a sufficiently small amount of investment are allowed to claim that cost as an expense for the year in which the investment took place, rather than over time. Over the past decade, the maximum amount of investment that a firm could claim as a tax-deductible expense in the year the investment occurred has varied from a low of $25,000 (the level in 2003 and also the current level) to as much as $500,000 from 2010 through 2013.

According to the latest available data on how depreciation deductions are distributed across industries, the manufacturing sector accounted for 28 percent of total depreciation and expensing deductions—slightly higher than its share of gross output in the economy, as manufacturing on average uses more physical capital than other industries.[45] Using those data on depreciation and expensing deductions, together with estimates of federal tax expenditures made by the staff of the Joint Committee on Taxation, CBO estimates that the federal revenue loss from offering those accelerated deductions to manufacturing firms in 2012 amounted to about $10 billion.[46]

Policies for Changing the Tax Treatment of Private Investment

Both of the tax policies discussed above have attracted their share of proposals for changes that might promote more innovation. However, the effects of tax preferences on businesses' behavior are difficult to predict. In particular, the interaction of many provisions of the tax code makes the effect of a single provision of the tax code on a firm's behavior difficult to discern.

Tax Credit for Increasing R&D Expenditures. Advocates of changing the incremental R&D tax credit to increase innovation have suggested three main types of modifications:

- Renew the credit and make it permanent,
- Make the credit simpler and reduce the potential for so-called windfalls, and
- Make the credit more supportive of new technology.[47]

Renew the Credit and Make It Permanent. The Congress is currently considering several measures to renew the R&D credit, which expired in December 2013, including permanently extending it (unlike previous extensions that were temporary). A permanent renewal would avoid the perennial negotiations over its expiration. However, since it has always been renewed, industry has probably come to treat the credit as effectively permanent anyway and to incorporate it into financial planning.

Simplify and Modify the Credit. One approach would retain the alternative simplified credit and eliminate the regular credit. The Government Accountability Office (GAO) has suggested that, to reduce the potential for firms to receive subsidies for R&D they would have undertaken anyway (so-called windfalls), lawmakers should also consider increasing the base for the alternative simplified credit from the current 50 percent of the average for the three prior years to a minimum of 50 percent of the current year's expenses.[48] Alternatively, some advocates argue that the credit rate should be increased because some other nations offer more generous research tax credits and, in general, provide more generous tax treatment of private-sector R&D overall.[49]

Make the Credit More Relevant to the Current Role of Technology. Many of the rules governing the credit were written decades ago when technology and high-tech firms played a very different role in the economy. For example, the credit is nonrefundable, meaning that only firms with tax liabilities can benefit from it in a given year, although firms can carry it forward in hopes of offsetting tax liabilities in the future. The lack of refundability means that the credit is often unavailable to firms during their start-up period, when they need assistance the most—a situation that is especially common in high-tech industries, given the relatively large number of start-ups in that sector.[50] Some analysts have suggested making the credit refundable by size or allowing start-up firms to sell unused credits. Making the credit refundable, however, would increase the likelihood that revenue losses are attributed to R&D activities that would have occurred without the credit or to activities that do not truly reflect innovative pursuits.

Similarly, R&D for software development for firms' internal use is explicitly *not* covered by the R&D credit.[51] When the law was written, most such software was used for purposes such as internal accounting processes. However, now that companies often provide Internet services using internally developed software, such an exemption may be outmoded.

Encouraging Domestic Manufacturing Through the Tax Code

It is difficult to determine whether providing more federal support for manufacturing would increase innovation. Offering investment incentives for manufacturing overall might have only a modest effect on R&D and innovation if less R&D–intensive manufacturing firms frequently took advantage of them. Although the manufacturing sector spends disproportionately on R&D, almost two-thirds (63 percent) of that R&D is conducted by firms in two manufacturing industries— computers and electronics; and chemicals, including pharmaceuticals—that have much larger ratios of R&D to output than the sector average (see Figure 2-7). Some service industries, most notably the computer software industry, have also become significant sources of R&D spending.[52]

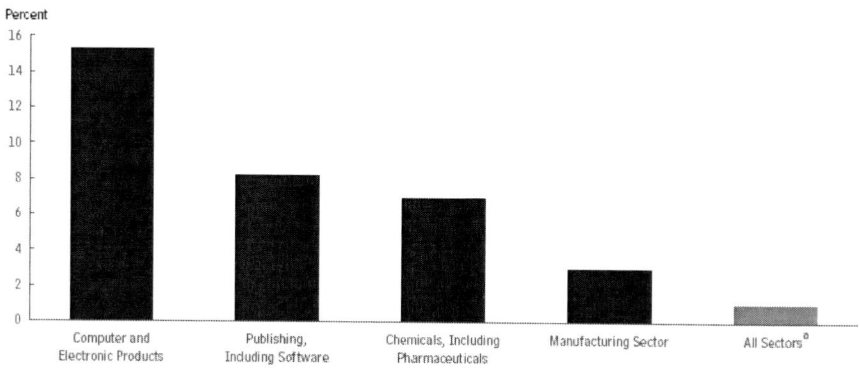

Source: Congressional Budget Office based on data from Raymond M. Wolfe, Business R&D Performance in the United States Increased in 2011, NSF Info Brief 13-335 (National Science Foundation, September 2013), Table 2, http://go.usa.gov/Be7J (PDF, 148 KB) and from the Bureau of Economic Analysis, Interactive Access to Industry Economic Accounts Data—GDP by Industry, "Gross Output by Industry" (January 23, 2014), www.bea.gov/iTable/index_industry_gdpIndy.cfm.

Note: R&D = research and development.

a. The category "All Sectors" does not include the agriculture, forestry, and fishing and hunting sector or the public administration sector.

Figure 2-7. Ratio of R&D Expenditures to Output, by Selected Industries and Sectors of the Economy, 2011.

Similarly, the argument that domestic firms—regardless of the industry they operate within—must have ready access to manufacturing capacity in order to innovate lacks broad evidence. For example, research on firms whose products are manufactured in one location—often a different country—while other parts of the business operate elsewhere, suggests that it may not be necessary to collocate manufacturing with product design and engineering.[53]

Proposals to increase support for manufacturing and innovation through the tax system include offering additional investment subsidies and increasing the value of other tax preferences for innovative activity. Such proposals include the following:

- Encouraging investment in R&D-related assets by expanding accelerated depreciation deductions, or extending bonus depreciation, for investments by firms in research-intensive industries, such as electronics and pharmaceuticals, and
- Offering preferential tax treatment for the sale or licensing of patented products (for example, allowing manufacturing firms to establish "patent boxes" that shelter a firm's patent-related profits).

The first proposal could be easier to implement than the second. Accurately determining the amount of a firm's profits that would be eligible to be sheltered in a patent box could be a problematic accounting exercise susceptible to misreporting. In addition, the prospect of offering preferential tax treatment for patented products could encourage firms to apply for a patent when they otherwise would not—for example, when the invention does not clearly meet the requirements for patentability—and thereby increase the number of low-quality patents and the costs to the economy associated with them (an issue that is discussed later in this report).

Federal Loan and Loan Guarantee Programs That Support Innovation

The federal government also supports innovation through credit programs that provide loans and loan guarantees to finance investments in innovative products or services. A limited number of federal credit programs provide loans or loan guarantees to private firms that bring new technologies to the market. However, those federal credit programs typically promote new technologies in the context of pursuing other policy goals, such as fostering renewable-energy technology to reduce carbon emissions. Increasing financial support for such federal credit programs may be difficult without funding projects that could obtain private funding anyway or whose social costs outweigh their benefits.

Current Federal Loan and Loan Guarantee Programs That Support Innovation

Federal credit programs that attempt to promote innovation rarely target innovation for its own sake; they typically promote new technologies in the context of pursuing other goals. Some programs increase the demand for existing high-tech products, such as airplanes or broadband equipment, by promoting exports or subsidizing their use in rural areas. The Small Business Administration provides loan guarantees ($2 billion in 2012) to Small Business Investment Companies, which in turn invest in small businesses, including some high-tech companies. The loans and loan guarantees that DOE has provided to businesses are perhaps the most prominent examples of the federal government's efforts to promote the deployment of new energy technologies through credit. The Administration has recently revived, after a hiatus of several years, three major credit programs administered by DOE that had issued more than $18 billion in loans and loan guarantees as of the end of fiscal year 2013:

- The Advanced Technology Vehicle Manufacturing (ATVM) program—a permanent loan program established in 2007 that aims to improve the energy efficiency of automobiles;
- The Section 1705 program—a temporary loan guarantee program authorized in 2009 that supported loans for some renewable-energy systems, electric power transmission, and innovative biofuel projects; and
- The Section 1703 program—a permanent loan guarantee program created in 2005 that aims to increase investment in nuclear facilities or other innovative energy facilities.

Roughly $4 billion remains available from prior appropriations to cover the subsidy cost of new loans under the ATVM program. Although the Section 1705 program expired on September 30, 2011, DOE has $170 million available for the subsidy costs of loan guarantees for certain projects under Section 1703. The Section 1703 program also is authorized to guarantee another $27.8 billion in debt, provided that recipients pay a fee covering the projected subsidy cost of those loans.[54]

To reduce their impact on the federal budget, the federal credit programs that support innovation often require the recipients of loans and guarantees to pay for the expected costs of that credit, which include losses from defaults net of any recoveries on the loans and guarantees.[55] Federal credit programs operate under the rules established by the Federal Credit Reform Act of 1990 for calculating the budgetary cost of direct loans and loan guarantees issued by the federal government. In general, before an agency can make loans or loan guarantees, lawmakers must provide funding sufficient to cover the government's cost, referred to as the subsidy cost, less fees paid by borrowers. Lawmakers control the amount of federal credit assistance either by appropriating the amount needed for the subsidies or, in cases in which net subsidy costs are negative, by setting limits on the volume of loans or loan guarantees. (As noted above in connection with student loans, the budgetary cost calculations established by the Federal Credit Reform Act do not provide the most comprehensive estimate of the costs of the loans or loan guarantees.)

Policies to Expand Federal Loan and Loan Guarantee Programs That Support Innovation

The rationale for increasing federal credit assistance to promote innovation is that young, innovative firms may lack adequate access to capital markets and face difficulties in getting private financing during the startup phase. When a firm decides to transform its research and development efforts into a commercial product, little funding—federal or otherwise—may be available for the process of bringing a product to market.

However, investors and institutions already devote substantial resources to bring innovative new products and services to market throughout the U.S. economy— in 2012, an amount equivalent to between $30 billion and $45 billion (in 2013 dollars)—undercutting arguments about a lack of capital.[56] Federal efforts (largely in R&D, not in credit activities) account for one-fifth to one-quarter of those figures.

Providing federal credit to promote innovation, such as new energy technology, without funding projects that could obtain private funding anyway or whose social costs outweigh their benefits, has proved challenging. For example, projects funded by DOE now account for a small fraction of U.S. solar capacity, with the great majority of capacity funded by sources other than DOE. The projects backed by DOE typically had secure streams of income in the form of power purchase agreements, usually from utilities promising to buy the energy that the projects produced.[57] Those factors suggest that much of the credit activity undertaken by DOE was replacing credit that could have been provided by the private sector. (However, the general tightening of credit after the fiscal year spanning 2008 through 2009—attributable both to market forces and to increased credit regulation— reduced the availability of credit for solar projects, so DOE may have supported some projects that could not have been undertaken otherwise.) Furthermore, much of the credit offered by DOE went to large companies, which are typically capable of accessing sufficient credit, rather than to start-ups, which are more likely to have had problems obtaining credit.[58]

Section 3. Altering the Legal and Regulatory Framework to Spur Innovation

To further encourage innovative activity in the United States, lawmakers could modify the legal and regulatory framework in which such activity takes place. For instance, foreign-born workers have long made the U.S. economy more innovative; therefore, easing the process by which highly skilled noncitizens are allowed to legally enter the country and secure employment could be expected to help boost innovation. Patent laws are meant to encourage innovation by allowing firms to capture more of the benefits of their investment in research and development; however, in spite of recent patent reform legislation passed by the Congress, persistent strains on the patent system suggest that additional measures may be needed to strengthen the link between patenting and innovation. And federal regulations governing the ways in which industries' products and services are introduced into the economy can affect their cost and the pace of their development.

Immigration of Highly Skilled Workers

Economic studies have generally found that increased immigration of foreign-born workers—and in particular those who are highly skilled—fosters innovation and productivity growth in the United States. However, U.S. immigration policy is more oriented toward family reunification than admitting workers who would be likely to contribute to innovation. Skilled workers represent only a modest share of the flow of the foreign born into the United States, receiving fewer temporary ("nonimmigrant") visas and permanent resident cards ("immigrant visas") than other groups of the foreign born receive. Recently, some policymakers have proposed changing immigration policy to increase the number of highly skilled noncitizens that are allowed to work in this country.

The Contribution of Skilled Immigrants to Innovation

Evidence suggests that foreign-born workers contribute disproportionately to innovation. In 2010, the foreign-born population accounted for 13 percent of the population of the United States.[1] Yet, in that same year, they accounted for 27 percent of all full-time workers specializing in the fields of science and engineering in the United States and 43 percent of all workers in this country with doctorates in science and engineering.[2] (At the same time, the foreign born are also disproportionately represented at the low-skill end of the workforce.)

Some of the contributions that certain immigrants make to innovation are related to their high level of educational attainment. High-tech firms, research hospitals, and universities, which account for much of the innovation in the economy, hire a disproportionate number of skilled foreign-born workers, and skilled workers in those institutions tend to invent and patent more than most other types of workers.[3]

But the presence of highly skilled foreign-born workers in high-tech industries does not tell the complete story: Foreign-born scientists also account for a disproportionate number of the star researchers in science. That is important because those exceptional individuals account for many of the big discoveries and conceptual breakthroughs that drive science. One team of science policy analysts, using a variety of measures—most highly cited 250 authors by field, highly cited papers and patents, and membership in the National Academies of Science and Engineering—concluded: "Although there is some variation by discipline, individuals making exceptional contributions to S&E [Science and Engineering] in the U.S. are disproportionately drawn from the foreign born."[4]

In an analysis of the effects of enacting S. 744, the Border Security, Economic Opportunity, and Immigration Modernization Act, the Congressional Budget Office concluded that its enactment would lead to higher productivity of both labor and capital because the increase in immigration—particularly of highly skilled immigrants—would tend to generate additional technological advancements, such as new inventions and improvements in production processes. CBO estimated that total factor productivity would be higher under that act by roughly 0.7 percent in 2023 and by roughly 1.0 percent in 2033, compared with what is projected to occur under current law.[5]

Policies Regarding the Immigration of Highly Skilled Workers

Highly skilled noncitizens in the workforce fall into one of three categories: lawful permanent residents on the path to U.S. citizenship ("immigrants"); legal temporary residents here for a limited time using one of several types of visa ("nonimmigrants"); and people working here without authorization. (Authorized *visitors*, such as tourists, are not counted in the foreign-born working population.) In addition to workforce members, there are also noncitizens studying in U.S. schools under student visas.

In 2013, U.S. authorities issued almost a million immigrant visas (commonly referred to as green cards). Almost two-thirds of those visas were issued under family-sponsored preference policies. Current policy limits the number of new employment-based immigrant visas to 140,000 plus the difference between the number of family-sponsored immigrant visas issued in the previous fiscal year and the cap on such visas. As a result, only 160,000 immigrant visas issued in 2013 were related to employment; most of those were issued to people with advanced degrees or special skills.[6] About half of the recipients already lived in the United States when their visa was issued.

The most prominent employment-based visas that admit highly skilled workers on a temporary basis are the H-1B and L-1 visas, which are work permits issued to specific employers.[7] Whereas H-1B visas are issued to permit the hiring of new employees, L-1 visas authorize intracompany transfers. In 2013, the Department of State issued 153,000 H-1B and 67,000 L-1 visas.[8] (The statutory cap on H-1B visas is 65,000, but a substantial number of exclusions apply.) Visas are valid for several years but can be extended. Sometimes they are not used immediately, but rather banked by the companies. Consequently, there is no official count of the actual number of workers holding such temporary work permits. One recent estimate suggested that there are roughly 1 million holders of H-1B or L-1 visas in the United States.[9]

Three types of institutions use those employment-based visas extensively: nonprofit research organizations, such as universities and research hospitals; U.S.-based multinational corporations; and offshore firms with a U.S. presence. H-1B visas issued to nonprofit research institutions are not subject to the same numerical quotas as those issued to private firms and are typically used to employ postdoctoral graduates and other skilled personnel to do research. Given the substantial role played by postdoctoral graduates in conducting and publishing scientific research, the importance of such visas to innovation is clear. Offshore firms often use temporary employment-based visas to educate their workforce and to transfer U.S. expertise to their home country.[10] Congruent with that strategy, the largest such users of H-1B and L-1 visas rarely sponsor their temporary workers for U.S. immigration visas. After those workers return to other countries, the productivity of the workforce abroad might be enhanced by their preceding time in this country but U.S. productivity might not be affected very much. However, while working in the United States for such offshore firms, workers holding H-1B visas are presumably productive members of the U.S. labor force, making contributions as other employees do. The U.S.-based multinational firms are more varied in their use of H-1B and L-1 visas.

Roughly one million foreign-born students were enrolled in certified academic or vocational programs in July 2014.[11] Arriving in the United States under a student visa and eventually becoming a permanent resident is a major path to immigration for highly skilled workers.

Effects of the Immigration of Highly Skilled Workers on Domestic Wages and Employment

Analysts disagree as to the effects of the immigration of highly skilled workers on the employment and wages of similarly skilled native-born workers, but generally the effects are believed to be small.[12] Some studies find that highly skilled native-born workers suffer job losses or reduced wages when faced with competition from highly skilled immigrants, whereas other studies—addressing different circumstances and using different data sets—find that an increase in the immigration of highly skilled workers increases the demand for skilled native-born employees.[13] In the face of an increase in the supply of highly skilled, foreign-born workers, native-born scientists and other workers in the fields of science, technology, engineering, and mathematics may shift to jobs and occupations (such as management and the sale of services and goods produced in STEM industries) in which their better command of the language and customs provides them with a comparative advantage, leaving the more technical aspects to skilled immigrants.[14] Native-born workers who, in terms of skills and inclination, are the closest substitutes for immigrants may have opportunities diminished by the immigration of talented scientists and other STEM workers. In its analysis of the effects of enacting S. 744, CBO estimated that relative wages would decrease slightly over several years for workers in the top quintile of the skill distribution relative to the wages that would occur under current law.[15]

Increasing the employment of young, highly skilled immigrants does seem to decrease the employment of older native-born workers, either absolutely or as a share of total employment, depending on the study.[16]

Policies to Increase the Immigration of Highly Skilled Workers

Policymakers have proposed changing immigration law in different ways to increase the immigration of highly skilled workers, including the following:

- Issuing more H-1B visas for foreign-born workers,
- Increasing immigrant visas for foreign-born workers, and
- Allowing foreign students in STEM fields to stay in the United States after graduation, either temporarily or permanently.

Issue More H-1B Visas. Eliminating or raising the quotas for H-1B temporary workers would increase the number of highly skilled workers in the United States. Such a change might also provide skilled noncitizens with an easier or quicker path to working in the United States than immigrant visas because of the current backlog in issuing immigrant visas.

However, some observers raise concerns that this approach would not be the best way to increase the immigration of highly skilled workers. Adding more visas for temporary workers could increase the pressure for the government to issue immigrant visas in the long run as workers already in the country seek to gain permanent resident status. And some view the H-1B system as flawed because it gives employers more control over workers as those workers become dependent on their employers to sponsor or renew their visas, which can keep wages low. Finally, the institutions using H-1B visas that are most likely to use them to further research— universities, research hospitals, and other nonprofit research organizations—are not bound by the current H-1B limits and so would not benefit from increases in H-1B quotas.

Increase Immigrant Visas for Highly Skilled Workers. Policymakers could also increase the number of immigrant visas for skilled workers. That would allow more highly skilled immigrants to stay in the country permanently, although that approach runs counter to the long-standing family-oriented focus of U.S. immigration policy. The processes that employers must generally undertake before seeking to hire a foreign-born worker (on either an immigrant visa or nonimmigrant visa) are designed to ensure that no American worker is displaced and that hiring a foreign-born worker does not permit an employer to depress wages or work conditions—but some researchers question how effective those processes are in achieving those goals.

Grant Foreign STEM Students Visas Upon Graduation. This approach would allow foreign students in STEM fields to stay in the United States after graduation, either temporarily or permanently. Under the current system, many students who specialize in STEM-related fields do remain in this country after graduation. According to one study, 64 percent of foreign recipients of a Ph.D. in a STEM-related field were still in the United States five years after graduation. Foreign-born workers with Ph.D.s in computer science and electrical engineering remained in the United States at even higher rates.[17]

Providing temporary employment-based visas for foreign students specializing in STEM fields would make it much easier for those students to remain in the United States after they graduate.[18] Immigrant visas for STEM students, for example, could take the form of granting lawful permanent resident status, commonly known as a green card, for graduates in STEM fields. Linking education in the United States to immigration prospects would affect foreign students' incentives for obtaining such a degree and U.S. universities' incentives to provide them to foreigners. In some countries that have tried such policies, certain educational institutions have increased their foreign enrollment substantially while lowering their educational standards.[19]

The Patent System

The patent system contributes to innovation and economic growth in two principal ways. First, it provides an incentive for innovative activity by enabling inventors— and those who commercialize an invention through a patent transfer or license—to make money from their efforts by excluding others from applying the technical solution in question. Hence, the patent system effectively conveys a monopoly to the patent owner that can be very profitable.[20] Second, the patent system provides for the dissemination of useful technical knowledge by requiring that a patent be published, rather than kept as a trade secret. Innovative products like computers and pharmaceuticals are covered under the patent system.

However, problems with the patent system—including too many low-quality patents, the considerable length of time required to process patent applications, and the rising cost of patent infringement litigation—may have weakened the linkage between patenting and innovation. A variety of proposals have been made for improving the patent system, ranging from increasing the resources available to the U.S. Patent and Trademark Office (USPTO) and to the federal courts that adjudicate patent disputes, to making further changes to the USPTO's administrative practices and to patent law in general.

Identifying policies that would clearly make the patent system more conducive to innovation is challenging. Efforts to address one problem could exacerbate others or weaken the incentive to innovate, and research on the extent of those problems and their impact on innovation is limited or still in its early stages.[21]

Concerns About the Patent System

Those arguing for reform of the patent system claim that it makes little contribution to innovation and economic growth outside of a few industries (such as pharmaceuticals and life sciences) and, in some cases, may actually be an impediment. They emphasize poor patent quality, the increase in the time that patent applicants must wait for a decision about their application, and rising litigation and settlement costs associated with patent infringement suits. Such concerns contributed to the passage of patent reform legislation in 2011—the Leahy-Smith America Invents Act (AIA)—and they persist today.

Although domestic patenting by U.S. citizens and foreigners combined grew at about the same annualized rate as TFP from 1963 to 1983, the rate of increase in patent applications and grants is much higher than the growth in TFP since then (see Figure 3-1). This suggests that the large increase in patenting activity since 1983 may have made little contribution to innovation.[22]

Many critics of the patent system today argue that an overburdened USPTO has made it too easy to obtain a patent. As a result, patents are often of low quality— meeting only the lowest standards of usefulness, novelty, and "nonobviousness"— yet still allow an inventor to exclude others from using the patented technology. The claims of a low-quality patent also may not be drafted clearly, stating the invention in an overly broad way that encompasses many applications. In addition, a low-quality patent may describe an invention too vaguely for the disclosure of the technical knowledge to be informative to practitioners.[23]

The proliferation of low-quality patents is especially likely to result in "patent thickets," groups of patents that are relevant to developing a particular technology, which can slow the introduction of new products by making it time-consuming and costly to obtain licenses for all of the necessary patents. Smaller innovating firms also tend to suffer disproportionately because patent thickets provide a greater competitive advantage to the typically large firms that hold more patents than those firms would realize solely on the basis of their innovativeness.[24]

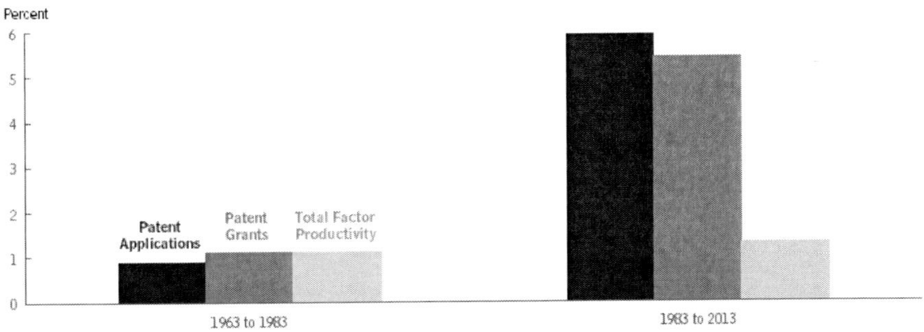

Source: Congressional Budget Office based on U.S. Patent and Trademark Office, Patent Technology Monitoring Team, "U.S. Patent Statistics Chart Calendar Years 1963–2013" (last modified July 24, 2014), http://go.usa.gov/ VnbC.

Note: Total factor productivity is the growth in output that remains after removing the contributions from growth in labor hours and capital services. Capital services are a measure of the flow of services available for production from the real (inflation-adjusted) stock of capital (typically, the physical structures and equipment used in production, along with intangible capital, such as computer software and knowledge gained from spending for research and development).

Figure 3-1. Annualized Growth in Patenting Activity and Total Factor Productivity, 1963 to 2013.

Another common cause for concern about the patent system is the increased time required for the USPTO to process patent applications. Patents can be an important way for start-up firms—particularly in the life-science industries—to signal to capital markets that they offer innovative products and, as a result, that they are a good investment. Delays in patent processing can deprive such firms of external funding. The amount of time that patent applicants had to wait for a decision about their applications increased from 19 months in 1992 to 29 months in 2013 (down from a high of over 35 months in

2010), with the number of pending patent applications increasing from approximately 270,000 to over 1.1 million.[25]

The patent system has also seen an increase in litigation and associated costs over the years. When a patent makes very ambiguous or broad claims to knowledge, it may be infringed upon inadvertently by many practitioners, which may allow the patent holder to assert its monopoly more broadly than would otherwise be warranted. The number of defendants in patent infringement lawsuits rose by 129 percent from 2007 through 2011, with lawsuits involving software-related patents (including patents on business methods) accounting for almost nine-tenths of that increase.[26]

Of particular concern are companies that accumulate patents without intending to apply that knowledge to the production of goods or services. Although defenders of such firms—referred to as patent assertion entities, or PAEs—argue that they provide a way for individual inventors to assert their patents more effectively than they could do on their own, critics emphasize that PAEs have an inherent advantage in patent litigation because they make no products and, as a result, are not vulnerable to infringement countersuits.[27] Evidence on the economic impact of PAEs remains sparse, but some studies have found that litigation involving those firms is very expensive and that inventors benefit relatively little from it. Research has also found that PAE lawsuits, or the threat of such litigation, impose a substantial burden on start-up firms (especially in information technology industries). Moreover, the patents asserted by PAEs often apply to products that small firms have simply purchased and are using in their business operations—rather than on patented technology that those firms have replicated without a license from the patent owner.[28]

Policies for Improving the Patent System

To address concerns about the quality of patents, reduce the length of time to process patent applications, and lower the cost and frequency of patent litigation, lawmakers could provide more resources to the patent system or make further changes to the USPTO's operational methods and to patent law beyond those put in place by the AIA.

Make More Resources Available. Under current law, the USPTO is funded by fees collected for the services it provides, such as patent application examinations and postgrant reviews. Reforms enacted in the AIA give the USPTO greater flexibility in setting those fees, thereby allowing the agency to align its fees more closely with its costs. However, the USPTO requires an appropriation from the Congress in order to spend the fees that it collects. Thus, USPTO's funding is considered to be discretionary and as such is limited by the caps on discretionary funding imposed by the Budget Control Act of 2011. Under those caps, the USPTO's funding has declined.

The Congress could increase the resources available to the patent system by making the revenues the USPTO collects independent of the appropriation process, by providing additional funding to the USPTO (say, by supplementing user fees with appropriations from the Treasury's general fund), by providing additional funding to the federal courts for handling patent cases, or through a combination of those actions.

Some observers argue that the USPTO should be granted additional appropriations because its need for more patent examiners, judges specializing in administrative law, and computer hardware and software exceeds what that agency could generate through a revised fee structure. Some observers also argue that if additional appropriations were provided to create more federal judgeships, the spreading of the existing workload across a larger number of judges would mean that patent lawsuits would be concluded more quickly and at less cost to litigants.[29] Some newly created judgeships could be for specialized patent litigation courts, set up along the same lines that bankruptcy courts are currently.

Improvements in the timeliness of patent decisions and legal rulings for patent enforcement could lead to greater demand for patents, to an extent limiting the improvements from providing further resources to the USPTO and federal courts. However, using more funds to increase the number of patent examiners could raise the quality of patents that are granted and thereby reduce low-quality patent applications by making it less likely that a patent would be awarded in such cases.

Change the Patent System. It is difficult to evaluate how proposals to change the patent system would ultimately affect the problems attributed to the patent system or its ability to better foster innovation. Efforts to address one problem could make other problems worse or reduce the incentive to innovate. For example, proposals that make it more costly to obtain and renew a patent or that weaken some of the protections enjoyed by patent owners might reduce the burdens on the patent system— thereby allowing patents to be processed more quickly, and litigation costs to fall—but could diminish the financial incentive to obtain a patent. Furthermore, research on a number of challenges to the patent system, such as those posed by the increase in litigation attributed to PAEs, is limited or still in its early stages.[30]

Policies to Reduce the Number of Low-Quality Patents. A number of changes have been suggested that aim to reduce the number of low-quality patents by modifying how the USPTO administers patent law:

- Charging higher fees to apply for or maintain a patent in order to discourage patents that have little economic value;
- Requiring that applications be reviewed by two examiners, independently, before a patent is granted;
- Making it more difficult or more costly to resubmit an application that was rejected; and
- Requiring the technical specifications of an invention to be described more precisely so that others may replicate the claimed invention.

Although increasing the scrutiny given to patent applications would reduce the number of low-quality patents that are granted, it might at the same time deter patent applications more broadly and could lengthen the processing time required for applications.[31] Currently, patent applications that have been rejected can be resubmitted— with only minor changes—for a new examination. Making it more difficult to resubmit a rejected application could reduce the processing time for the average patent application because such submissions (or "requests for continued examination") have been found to impose a significant burden on the USPTO. However, making it more difficult to obtain a patent could again negatively affect the incentive to patent.[32]

Policies to Reduce the Cost to Defendants of Infringement Litigation and to Reduce the Frequency of Nuisance Lawsuits. Most proposed changes to patent law are intended to reduce the cost to defendants of patent infringement lawsuits and the frequency of nuisance lawsuits. Those changes could involve reducing the potential payoff from engaging in patent litigation, making it harder for plaintiffs to prevail in patent infringement lawsuits, and limiting the scenarios in which patent infringement can be asserted.

Within that group, one set of proposals is focused more on the potential costs that defendants may bear as a result of going to court. That set of proposals includes the following measures: establishing legal protections for end users of products that embody technologies subject to a patent dispute; implementing "fee-shifting" in patent infringement lawsuits—and in particular, those initiated by PAEs—by requiring that the plaintiffs be required to pay all court costs in the event that they lose; and requiring that the damages awarded upon a finding of patent infringement be apportioned according to the contribution of the technology that is covered by the infringed patent or patents to the price of the good or service in question. However, the effects of the fee-shifting may be limited to the extent that some PAEs might engage in a nominal amount of production associated with a patent to be asserted in litigation so that they would not be considered a PAE. In addition, because a given product can encompass one or more technologies on which a number of patents have been granted, it could be very difficult to determine accurately how much of the value of that product was attributable to a small subset of those technologies.[33]

Another set of proposals is focused more on the likelihood that a patent-owner will prevail in a patent infringement lawsuit and includes the following measures: lowering the legal standard for finding that a patent is invalid (from the current "clear and convincing" threshold to a less stringent requirement of a "preponderance of the evidence") and, perhaps most controversially, establishing an "independent inventor defense" that would provide a safe harbor for users of a patented invention if they could prove that they did not copy the patent or otherwise learn about the technique from the patent owner. Some argue that lowering the threshold for finding a patent invalid would align the criteria that judges and juries use to determine whether a patent is valid with the diminished quality of patents that are currently granted. Allowing those accused of patent infringement to make an independent inventor defense would address research findings that the outright copying of a patent is seldom claimed in patent infringement lawsuits. However, both proposals would represent profound changes to patent law and could potentially dampen incentives to innovate.[34]

Still another set of changes that the Congress could make to patent law would be to modify patent law for particular kinds of inventions. Such changes would address the concern that the current unitary patent system, in which the same basic protections apply regardless of the type of invention, might not be appropriate for a variety of industries when the nature of technology development can be industry specific. Although patent protection is widely understood to provide an important incentive for the development of new drugs—which is typically very costly—the contribution of patents to innovation in software or business methods is often questioned because the costs of developing such new products and processes may be modest. One possible change to patent law that could reduce the cost and frequency of litigation would be to limit patent protections for inventions that were relatively inexpensive to develop. For example, patents on software and business methods could expire sooner than is the case today (which, with renewals, is after 20 years), reducing the incentive to obtain those patents.

Another change that could address patent quality, the processing burden on the USPTO, and the cost and frequency of litigation would be to limit the ability to obtain a patent on certain inventions.

Critics of proposals to move away from a unitary patent system argue that the benefits of such a switch would be limited by efforts on the part of patent applicants to thwart industry-specific restrictions on patent protections by stating their patent claims in such a way as to have them classified with inventions that qualify for the strongest protections (indeed, because there is no one category for "software" in the USPTO's patent classification scheme, researchers have developed several alternative categories that overlap in varying degrees). They also argue that making some types of inventions ineligible for a patent would not be permissible under the intellectual property treaties that the United States has signed with other countries. However, some legal scholars have questioned whether the United States is prohibited under international treaties from deviating in any way from a unitary patent standard (for example, drug manufacturers benefit from an exemption from patent infringement when they conduct clinical trials to establish the bioequivalence of a generic drug with a patented pharmaceutical).[35]

One consideration that surrounds a number of the modifications suggested for the patent system is whether actions already taken by the USPTO and federal courts will resolve those concerns. For example, some observers argue that the reforms implemented by the AIA have not yet had their full effect on the patent system, and so enacting additional measures in the near future would be premature. Beyond the provisions of the AIA, the USPTO has joined with the private sector to improve the quality of software patents (the Software Partnership) and is also implementing a number of Executive Orders to address other concerns about the patent system. In addition, recent court decisions may mitigate, if not make unnecessary, the need for the Congress to legislate reform in those areas.[36]

REGULATORY GOALS AND TOOLS

Lawmakers could enact legislation modifying regulations that affect the introduction of new products and services in order to promote innovation. For example, the Congress could reconsider the federal government's regulatory goals and reassess the way certain risks are balanced against the potential benefits provided by innovation. Separately, the Congress could change the tools of regulatory policy in some areas to allow for more flexibility in meeting regulatory goals, which would reduce the costs imposed by regulation and promote more innovation. Finally, in some instances, federal lawmakers could address the balance of risks and rewards codified by state and local governments.

Modifying Regulatory Goals

Society's decisions about how to regulate new products and services generally involve balancing the risks and rewards associated with those innovations. Regulation that aims to diminish the risks that accompany innovations can restrict innovations that have the potential to bring substantial benefits to society. In addition, regulation can deter innovation by imposing administrative costs associated with compliance, by creating uncertainty and delays, and by redirecting managerial attention from innovative activity to meeting regulatory requirements. Regulation may also create incentives for firms to divert resources to less promising investments to avoid the regulation.

Individual regulatory agencies are often limited in their ability to assess the full, long-term impact of regulations. Even if the cost of any individual regulation imposed by a single agency is small, the cumulative effect of a large body of regulations imposed by multiple agencies might be considerable. Furthermore, the ever-changing nature of technological innovation implies that regulations that appropriately balanced the risks and rewards of innovation in the past might now be outmoded, and new technology might open opportunities that were not considered when the regulations were designed. In such circumstances, it might be advantageous to reassess regulations on a recurring basis.

Regulation of financial markets is one important example of the trade-off between innovation and risk. Many financial innovations have greatly benefited consumers. For example, the use of credit cards and automated teller machines is now essential to the workings of the financial system. However, financial innovations also create risks. In the years leading up to the financial crisis, the enormous growth in nontraditional mortgages and mortgage-related securities and derivatives facilitated a sharp increase in leverage among households and financial institutions. That leverage increased the risk in the financial system and the economy, as became apparent during the financial crisis and severe recession. In response, policymakers have sought to mitigate the costs and risks generated by financial innovations through regulation.[37] Often, regulations put in place to reduce the risks that accompany financial innovations also diminish the profits that flow to the creators of those innovations. In turn, that reduction in profits dampens the incentive to innovate.

Some regulatory activity serves to limit investment or raises the cost of providing an existing product or service. For example, federal regulations to reduce emissions from coal-burning plants, industrial boilers, and some other industrial process heaters may require tens of billions of dollars of specialized investment by the affected companies.[38] In the absence of regulations requiring the firms to make such investments in pollution-control technology, the affected companies might make additional investments in productivity-enhancing equipment. However, the magnitude of any reduction in productivity-enhancing investment because of environmental regulations is unclear. More generally, the relationship between different types of investment and innovation is not straightforward. For example, some environmental requirements can be fulfilled only by introducing new technology.[39]

Adopting Flexible Regulatory Tools

In recent decades, the Congress has sometimes increased the flexibility of regulation by allowing firms to focus on what is to be achieved, such as safety or emission goals, rather than specifying how those goals are to be achieved. For example, instead of mandating that firms install certain pieces of equipment to reduce emissions, the 1990 amendments to the Clean Air Act created a market in tradable allowances that permitted firms to bank or trade the right to emit sulfur dioxide. Firms with low compliance costs could sell their allowances to firms with high compliance costs, reducing the aggregate costs of emissions control. Firms could also save their allowances to be used as needed to reduce their future compliance costs.[40]

However, such regulatory flexibility is still unevenly provided. Analysts point out that federal regulations are increasing in both number and complexity even as some agencies experiment with more flexible forms of regulation.[41] The number of restrictions or binding legal obligations imposed by federal regulations continues to rise. That "regulatory accumulation" can impose costs beyond the costs of individual regulations through interactions or by overburdening the management of affected companies with regulatory compliance. Even the successful trade in sulfur emission permits discussed above was largely brought to a halt after more than a decade in operation by other parts of the Clean Air Act.[42]

Several Administrations have also launched efforts to make regulation less costly to the economy. The current Administration has had a "look back" initiative under way since 2011, aimed at having regulatory agencies reconsider their existing rules and explore ways to reduce the number of regulatory requirements or to make them less costly to implement.[43] Nevertheless, at present, the process of reviewing existing regulations is less well institutionalized than the process of examining regulations prospectively, as agencies receive little detailed guidance on how to reconsider regulations that are currently in effect. Some analysts have suggested creating a formal framework for evaluating those regulations regularly, either in established agencies or by creating a new one for that purpose.

Influencing State and Local Regulation

Federal policymakers could also act when other regulatory actors in the economy place a lower value on the rewards of innovation and a higher value on its risks than do federal lawmakers. For example, in 1986, the Congress passed the National Childhood Vaccine Injury Act, which was designed, among other things, to reduce the risk of liability associated with the development and sale of vaccines. Product liability is typically governed by state law, but that act, which created a federal alternative dispute resolution mechanism for drug companies and aggrieved individuals, has been generally regarded as a success in encouraging the development of new vaccines.[44]

Policymakers could extend that experience in restricting liability rules into new areas in which innovation is likely to be inhibited by liability laws, such as the development of self-driving automobiles. Advocates argue that the widespread use of such automobiles would reduce the number of traffic fatalities, which currently stand at 33,000 deaths per year, a substantial portion of which are caused by drunken or distracted drivers.[45] Yet, despite those automated vehicles having the potential to improve national traffic safety, rules regarding strict liability for new products could cause the producers of the first self-driving cars to be judged not against the 33,000 actual deaths, but against an ideal of no deaths. If policymakers wanted to set a different standard, they could revisit the balance between liability and innovation for this nascent technology. Other analysts argue that federal preemption is not necessary: State liability law has been regularly interpreted by state courts to accommodate new technology.[46] Alternatively, federal policymakers could help create appropriate liability insurance markets—much as they did with reinsurance against terrorist attacks after 2001.

Even as the Congress considers how to best balance innovation against other policy goals, the federal system of government places limits on its jurisdiction. For example, in the case of broadband transmission facilities, policies of local governments and regulatory authorities—local zoning regulations and access to local infrastructure such as utility poles and rights of way—are important determinants of the pace of the expansion. The Congress, most recently through the Middle Class Tax Relief and Job Creation Act of 2012, and the Federal Communications Commission have both placed certain limitations on local zoning authorities' regulation of the placement of broadband equipment and facilities without totally preempting state and local law.[47] In this case and others, federal agencies can influence the policies of local government and regulatory authorities only slowly and with some difficulty.

End Notes for Section 1

[1] With its latest comprehensive revision to the national income and product accounts (released in July 2013), the Bureau of Economic Analysis (BEA) began classifying as investment R&D spending by government, businesses, and nonprofit organizations that serve households (together with business spending on some other types of intellectual property, such as music, movies, and other original artistic and literary works). BEA also revised its historical estimates of the capital stock to incorporate that reclassification (back to 1959 for business spending on R&D and even earlier for works of entertainment, such as music and books). A comparison of the sources of economic growth before and after that reclassification shows little qualitative change in the contribution of TFP. Some researchers suggest that spending by firms on certain other activities also results in innovation and influences TFP; examples include spending on industrial design, worker training, and firm branding and marketing. See Carol A. Corrado and Charles R. Hulten, "How Do You Measure a 'Technological Revolution'?" American Economic Review, vol.100, no. 2 (May 2010), pp. 99–104, http://tinyurl.com/nb6vs7z.

[2] Congressional Budget Office, Changes in the Distribution of Workers' Hourly Wages Between 1979 and 2009 (February 2011), pp. 6–10, www.cbo.gov/publication/22010.

End Notes for Section 2

[1] Most agencies disburse their annual appropriations over more than one fiscal year—consequently, there may be discrepancies between funding (appropriations) and spending (outlays) in any given year. In addition, the estimates given here are for the conduct of R&D only. Although most federal spending for R&D involves the actual conduct of R&D, federal agencies also spend a small fraction of their R&D funds on equipment and facilities.

[2] National Science Foundation, "National Patterns of R&D Resources: 2011–12 Data Update" (December 2013), Table 1, http://go.usa.gov/EWBH.

[3] The Department of Energy and the Department of Homeland Security account for a small fraction of defense R&D.

[4] American Association for the Advancement of Science, Trends in Defense R&D, FY 1976-2015 (May 6, 2014), http://tinyurl.com/lh63smp (XLS, 14.6 KB).

[5] See Congressional Budget Office, Federal Investment (December 2013), www.cbo.gov/publication/44974; and Andrew A. Toole, "Does Public Scientific Research Complement Private Investment in Research and Development in the Pharmaceutical Industry?" Journal of Law and Economics, vol. 50, no. 1 (February 2007), pp. 81–104, http://dx.doi.org/10.1086/508314.

[6] The development of weapon systems primarily contributes to the national welfare through its role in military policy rather than through increases in economic productivity. Because weapons development is largely devoted to optimizing the operation of specific weapon systems, the results of that work generally have narrower applicability than does research in broader areas of science and technology. Reviews of the literature suggest that such optimization and other product development generally lead to fewer benefits for other parts of the economy than basic and applied research. See Bronwyn H. Hall, Jacques Mairesse, and Pierre Mohnen, Measuring the Returns to R&D, Working Paper 15622 (National Bureau of Economic Research, December 2009), pp. 19–20, www.nber.org/papers/w15622.

[7] Iain M. Cockburn and Rebecca M. Henderson, "Publicly Funded Science and the Productivity of the Pharmaceutical Industry," in Adam B. Jaffe, Josh Lerner, and Scott Stern, eds., vol. 1 of Innovation Policy and the Economy (National Bureau of Economic Research, January 2001), pp. 1–34, www.nber.org/chapters/c10775. For more information on life expectancy, see David M. Cutler and Mark McClellan, "Is Technological Change in Medicine Worth It?" Health Affairs, vol. 20, no. 5 (September 2001), pp.11–29, http://content.healthaffairs.org/content/20/5/11.full.

[8] Jack W. Scannell and others, "Diagnosing the Decline in Pharmaceutical R&D Efficiency," Nature Reviews Drug Discovery, vol. 11, no. 3 (March 2012), pp. 191–200, www.nature.com/nrd/journal/v11/n3/full/nrd3681.html.

[9] Nicholas Bloom, Mark Schankerman, and John Van Reenen, Identifying Technology Spillovers and Product Market Rivalry, Centre for Economic Performance Discussion Paper 675 (London School of Economics and Political Science, December 2012), pp. 3–4, http://tinyurl.com/Kmv4srm (PDF, 1 MB).

[10] Congressional Budget Office, Long-Term Implications of the 2014 Future Years Defense Program (November 2013), www.cbo.gov/publication/44683.

[11] Congressional Budget Office, Federal Investment (December 2013), p.12, www.cbo.gov/publication/44974.

[12] Congressional Budget Office, Federal Climate Change Programs: Funding History and Policy Issues (March 2010), pp. 16–19, www.cbo.gov/publication/21196. See also Peter Ogden, John Podesta, and John Deutch, "A New Strategy to Spur Energy Innovation," Issues in Science and Technology, vol. 24, no. 2 (Winter 2008), http://issues.org/24.2/ogden.html.

[13] Each year since 1963, R&D Magazine has recognized the 100 best inventions that are incorporated into commercial products. The survey in question tracked the sources of the inventions. Fred Block and Matthew R. Keller, Where Do Innovations Come From? Transformations in the U.S. National Innovation System, 1970–2006 (July 2008), http://tinyurl.com/mzxys7g. Federal laboratories include not just the major DOE weapons laboratories, but those of all agencies.

[14] Federal laboratories often have specialized equipment or expertise that help private firms develop technology. National Institute of Standards and Technology, Federal Laboratory Technology Transfer, Fiscal Year 2011: Summary Report to the President and the Congress (September 2013), p. 8, http://go.usa.gov/K5Qx. For laboratory count, see Federal Laboratory Consortium, "About the FLC" (accessed November 10, 2014), www.federallabs.org/flc/home/about/.

[15] For data on small business awards, see Small Business Innovation Research program, "Awards" (accessed May 15, 2014), www.sbir.gov/past-awards.

[16] Congressional Budget Office, Federal Efforts to Reduce the Cost of Capturing and Storing Carbon Dioxide (June 2012), www.cbo.gov/publication/43357.

[17] David J. Carney and Patricia Oberndorf, A Summary of DOD COTS-Related Policies (Carnegie Mellon Software Engineering Institute, September 1998), http://tinyurl. com/n4oax7m.

[18] See Congressional Budget Office, The Pell Grant Program: Recent Growth and Policy Options (September 2013), www.cbo.gov/publication/44448.

[19] National Center for Education Statistics, Digest of Education Statistics 2012 (January 2013), Table 202, http://go.usa.gov/7DDA. The federal share might have been exceptionally high in the year displayed because of spending authorized by the American Recovery and Reinvestment Act of 2009.

[20] Department of Education, FY 2014 Department of Education Justifications of Appropriation Estimates to the Congress, vol. 1, "Student Loans Overview," p. S-12, http://go.usa.gov/7SZR.

[21] Under procedures established by FCRA, the cost of a student loan is estimated in the federal budget by taking into account the amount of the loan, expected payments to the government over the life of the loan, and other cash flows—all discounted to a present value using interest rates on Treasury securities. See Congressional Budget Office, Options to Change Interest Rates and Other Terms on Student Loans (June 2013), www.cbo.gov/publication/44318. In contrast, the fair-value approach uses an estimate of the discount rates that private investors would use to discount the risky cash flows from student loans. The difference between the private discount rates and the rates paid on Treasury securities reflects the cost of market risk associated with those loans. (Market risk is the component of risk that remains even after a portfolio has been diversified as much as possible.) For additional discussion, see testimony of Douglas W. Elmendorf, Director, Congressional Budget Office, before the House Committee on Financial Services, Estimates of the Cost of the Credit Programs of the Export-Import Bank (June 2014), www.cbo.gov/publication/45468.

[22] Joint Committee on Taxation, Estimates of Federal Tax Expenditures for Fiscal Years 2014–2018, JCX-97-14 (August 2014), pp. 29–30, www.jct.gov/publications.html?func=select&id=5.

[23] Heather B. Gonzalez and Jeffrey J. Kuenzi, Science, Technology, Engineering, and Mathematics (STEM) Education: A Primer, Report for Congress R42642 (Congressional Research Service, January 2014).

[24] See Government Accountability Office, Strategic Planning Needed to Better Manage Overlapping Programs Across Multiple Agencies (January 2012), www.gao.gov/products/GAO-12-108; and National Technology and Science Council, The Federal Science, Technology, Engineering, and Mathematics (STEM) Education Portfolio (December 2011), p. xi, http://go.usa.gov/7v5H (PDF, 1.34 MB).

[25] Some undergraduates also work on such grants, but it is not generally a central component of their education.

[26] National Science Foundation, Science and Engineering Indicators 2014 (February 2014), Appendix Table 5-2, www.nsf.gov/statistics/seind14/index.cfm/appendix.

[27] Margaret E. Blume-Kohout and John W. Clack, "Are Graduate Students Rational? Evidence From the Market for Biomedical Scientists," PLOS ONE, vol. 8, no. 12 (December 2013), http://dx.doi.org/10.1371/journal.pone.0082759.

[28] Richard B. Freeman, Tanwin Chang, and Hanley Chiang, "Supporting 'The Best and Brightest' in Science and Engineering: NSF Graduate Research Fellowships," in Richard B. Freeman and Daniel Goroff, eds., Science and Engineering Careers in the United States: An Analysis of Markets and Employment (University of Chicago Press, 2009), pp. 19–57, http://papers.nber.org/books/free09-1.

[29] B. Lindsay Lowell and Harold Salzman, Into the Eye of the Storm: Assessing the Evidence on Science and Engineering Education, Quality, and Workforce Demand (Urban Institute, October 2007), p. 30, www.urban.org/publications/411562.html. Other analysts dispute those assertions. See Robert D. Atkinson and Luke A. Stewart, The Real Story on Guestworkers in the High-Skill U.S. Labor Market (Information Technology and Innovation Foundation, May 2013), http://tinyurl.com/q7db28p.

[30] National Science Foundation, Science and Engineering Indicators 2014 (February 2014), Appendix Table 3-3, www.nsf.gov/statistics/seind14/index.cfm/appendix.

[31] Jonathan Rothwell, The Hidden STEM Economy (Brookings Institution, June 2013), p. 7, http://tinyurl.com/nvvhrq5.

[32] Ibid., p. 19.

[33] Richard M. Ingersoll and David Perda, "Is the Supply of Mathematics and Science Teachers Sufficient?" American Educational Research Journal, vol. 43, no. 3 (September 2010), pp. 563–594, http://dx.doi.org/10.3102/0002831210370711.

[34] Organisation for Economic Co-operation and Development, Programme for International Student Assessment, "Country Note: United States" (2012), www.oecd.org/pisa/key findings/pisa-2012-results.htm.

[35] Department of Education, National Center for Education Statistics, Schools and Staffing Survey (SASS), Public School Teacher Data File, 2007–2008, http://go.usa.gov/7SnC.

[36] The tax code also favors R&D by allowing firms to deduct the cost of R&D in the year those expenses are incurred as opposed to depreciating them over time. Changing that permanent feature of the tax code is rarely proposed. For 2014, the Joint Committee on Taxation estimates that the expensing of private R&D cost the Treasury $4.7 billion. See Joint Committee on Taxation, Estimates of Federal Tax Expenditures for Fiscal Years 2014–2018, JCX-97-14 (August 2014), www.jct.gov/publications.html?func=select&id=5.

[37] Senate Committee on the Budget, Tax Expenditures: Compendium of Background Material on Individual Provisions, S. Prt. 112-45 (December 2012), pp. 95–111, http://go.usa.gov/A4dj(PDF, 53 MB).

[38] Joint Committee on Taxation, Estimates of Federal Tax Expenditures for Fiscal Years 2012–2017, JCS-1-13 (February 2013), http://go.usa.gov/KDKV.

[39] Laura Tyson and Greg Linden, The Corporate R&D Tax Credit and U.S. Innovation and Competitiveness: Gauging the Economic and Fiscal Effectiveness of the Credit (Center for American Progress, January 2012), Table 7, http://tinyurl.com/pvmd3r3.

[40] The credit also had two additional components: a basic research credit of 20 percent for the funding of university or other nonprofit research and an energy research credit of 20 percent

for research at qualified consortiums. A permanent tax credit also exists for research related to orphan drug development.

41 For a recent review of this literature, see Joint Committee on Taxation, Background and Present Law Relating to Manufacturing Activities Within the United States, JCX-61-12 (July 2012), pp. 119–121, http://go.usa.gov/KDQQ.

42 For example, before the credit became available, it did not matter whether an employee's salary was for conducting R&D or not. Either way, it was deducted as an ordinary business expense. With the credit, classifying salaries as related to R&D could be of additional value to the firm.

43 Government Accountability Office, Tax Policy: The Research Tax Credit's Design and Administration Can Be Improved, GAO-10-136 (November 2009), pp. 16–18, www.gao.gov/products/GAO-10-136.

44 See, for example, Gary P. Pisano and Willy C. Shih, "Restoring American Competitiveness," Harvard Business Review (July–August 2009), http://hbr.org/archive-toc/BR0907.

45 Joint Committee on Taxation, Background and Present Law Relating to Manufacturing Activities Within the United States, JCX-61-12 (July 2012), pp. 28–43 and 96–98, http://go.usa.gov/KDQQ.

46 Joint Committee on Taxation, Estimates of Federal Tax Expenditures for Fiscal Years 2012–2017, JCS-1-13 (February 1, 2013), p. 34, http://go.usa.gov/KDKV. Tax expenditure estimates of accelerated depreciation for equipment investment (which include the effects of bonus depreciation) must be interpreted carefully. For example, because the provisions regarding bonus depreciation were not extended after 2013, the staff of the Joint Committee on Taxation estimates that negative tax expenditures will occur between 2014 and 2016.

47 For a more extensive list of possible modifications, see Robert D. Atkinson, "Expanding the R&E Tax Credit to Drive Innovation, Competitiveness and Prosperity," Journal of Technology Transfer, vol. 32, no. 6 (December 2007), pp. 617–628, http://dx.doi.org/10.1007/s10961-007-9046-y.

48 Government Accountability Office, Tax Policy: The Research Tax Credit's Design and Administration Can Be Improved, GAO-10-136 (November 2009), p. 39, www.gao.gov/products/GAO-10-136.

49 Luke A. Stewart, Jacek Warda, and Robert D. Atkinson, We're #27: The United States Lags Far Behind in R&D Tax Incentive Generosity (Information Technology and Innovation Foundation, July 2012), http://tinyurl.com/opd5qdz.

50 Gary Guenther, Research Tax Credit: Current Law and Policy Issues for the 113th Congress, Report for Congress RL31181 (Congressional Research Service, February 2014), pp. 24–29.

51 Testimony of Annette Nellen, Professor, College of Business, San Jose State University, before the Senate Finance Committee, Tax Reform Options: Incentives for Innovation (September 20, 2011), http://go.usa.gov/KDhh (PDF, 365 KB).

52 A similar pattern of R&D intensity across those industries emerges when comparing R&D spending and the number of workers conducting R&D to industry net sales and employment, respectively; see Raymond M. Wolfe, Business R&D Performance in the United States Increased in 2011, NSF Info Brief 13-335 (National Science Foundation, September 2013), http://go.usa.gov/Be7J (PDF, 148 KB). The concentration of R&D spending in certain industries may arise because firms in those industries may be inherently more likely than others to be innovative, given both current technologies and demand for the goods and services they produce. That could explain why firms specializing in computers, electronics, and software are the most R&D–intensive and why the software

industry emphasizes R&D whereas other service-sector firms spend comparatively little relative to their gross output.

[53] See Andrew B. Bernard and Teresa C. Fort, Factoryless Goods Producers in the U.S., Working Paper 19396 (National Bureau of Economic Research, August 2013), www.nber.org/papers/w19396; and Kimberly Bayard, David Byrne, and Dominic Smith, The Scope of U.S. Factoryless Manufacturing (Upjohn Institute, February 2013), www.upjohn.org/MEG/papers/baybyrsmi.pdf (674 KB).

[54] For a discussion of factors that influence the credit risk of some of the projects eligible for assistance under the Section 1703 program, see Congressional Budget Office, Federal Loan Guarantees for the Construction of Nuclear Power Plants (August 2011), www.cbo.gov/publication/41510.

[55] Estimates of the risks of default, and the consequent budgetary costs, change over time as economic conditions evolve and government agencies gain more experience with their loans and loan guarantees. As a result, the estimated subsidy cost of federal loans and loan guarantees is frequently revised. Under the Federal Credit Reform Act, such revisions are recorded in the budget as "credit re-estimates" on an annual basis.

[56] A survey of corporate R&D activity estimated that the outer bounds of the amount of resources used for commercializing new products and services was between 2 percent and 14 percent of aggregate R&D spending in the economy. CBO estimates a likely range for those amounts in 2012 to be between $30 billion and $45 billion. See also Lewis M. Branscomb and Philip E. Auerswald, Between Invention and Innovation: An Analysis of Funding for Early-Stage Technology Development , NIST GCR 02-841 (National Institute of Standards and Technology, November 2002), p. 62, http://go.usa.gov/76Qk (PDF, 631 KB).

[57] Jeff St. John, "After a Hiatus, DOE's Clean Energy Loan Program Will Soon Be Back in Action," GreenTechMedia (February 27, 2014), http://tinyurl.com/qef4tqo. Unlike the utility purchase guarantees associated with producing solar electricity, DOE's loan guarantees for manufacturing solar equipment were not backed by such agreements.

[58] Testimony of Veronique de Rugy, Senior Research Fellow, George Mason University, before the House Committee on Oversight and Government Reform, Assessing the Department of Energy Loan Guarantee Program (June 19, 2012), http://tinyurl.com/qfneumu.

End Notes for Section 3

[1] Congressional Budget Office, A Description of the Immigrant Population—2013 Update (May 2013), Exhibit 1, www.cbo.gov/publication/44134.

[2] National Science Foundation, Science and Engineering Indicators 2014 (February 2014), Table 3-27, www.nsf.gov/statistics/seind14/.

[3] Jennifer Hunt and Marjolaine Gauthier-Loiselle, "How Much Does Immigration Boost Innovation?" American Economic Journal: Macroeconomics, vol. 2, no. 2 (April 2010), p. 52, http://dx.doi.org/10.1257/mac.2.2.31.

[4] Paula E. Stephan and Sharon G. Levin, "Exceptional Contributions to U.S. Science by the Foreign-Born and Foreign-Educated," Population Research and Policy Review, vol. 20, no. 1–2 (April 2001), pp. 59–79, http://dx.doi.org/10.1023/A:1010682017950.

[5] Congressional Budget Office, The Economic Impact of S. 744, the Border Security, Economic Opportunity, and Immigration ModernizationAct (June 2013), pp. 12–13, www.cbo.gov/publication/44346.

[6] Randall Monger and James Yankay, U.S. Lawful Permanent Residents: 2013 Annual Flow Report (Department of Homeland Security, May 2014), Table 2, http://go.usa.gov/76fd.

[7] Other, smaller visa programs cover people with certain extraordinary abilities and their assistants.

[8] Department of State, Report of the Visa Office 2013, "Nonimmigrant Visas Issued by Classification (Including Crewlist Visas and Border Crossing Cards): Fiscal Years 2009–2013," Table XVI (B), http://go.usa.gov/76GT.

[9] Ron Hira, Bridge to Immigration or Cheap Temporary Labor? The H-1B & L-1 Visa Programs Are a Source of Both, EPI Briefing Paper 257 (Economic Policy Institute, February 2010), p. 3, www.epi.org/publication/bp257/.

[10] Ibid.

[11] Immigration and Customs Enforcement, Student and Exchange Visitor Information System, "General Summary Quarterly Review: SEVIS by the Numbers" (July 2014), p. 8. www.ice.gov/sevis.

[12] Congressional Budget Office, The Economic Impact of S. 744, the Border Security, Economic Opportunity, and Immigration Modernization Act (June 2013), pp. 19–20, www.cbo.gov/publication/44346.

[13] For a review of the conflicting studies, see William R. Kerr, U.S. High-Skilled Immigration, Innovation, and Entrepreneurship: Empirical Approaches and Evidence, Working Paper 19377 (National Bureau of Economic Research, August 2013), pp. 9–17, www.nber.org/papers/w19377.

[14] Ibid., p. 16.

[15] Economic studies have generally found that increases in the number of skilled workers raise the productivity of less skilled workers. See Congressional Budget Office, The Economic Impact of S. 744, the Border Security, Economic Opportunity, and Immigration Modernization Act (June 2013), p. 20, www.cbo.gov/publication/44346.

[16] One analyst notes that substituting older native-born workers with younger foreign-born workers offers a firm two types of cost savings: using less-costly foreign workers instead of relatively expensive natives, and shifting away from older workers (who tend to command higher salaries) toward younger ones. See Norman Matloff, "On the Need for Reform of the H-1B Non-Immigrant Work Visa in Computer-Related Occupations," University of Michigan Journal of Law Reform, vol. 36, no. 4 (2003), p. 816. Furthermore, in STEM fields, there is a move away from older workers toward younger workers generally. Sari Pekkala Kerr, William R. Kerr, and William F. Lincoln, Skilled Immigration and the Employment Structures of U.S. Firms, Working Paper 19658 (National Bureau of Economic Research, November 2013), p. 4, www.nber.org/papers/w19658. For a more general review, see William R. Kerr, U.S. High-Skilled Immigration, Innovation, and Entrepreneurship: Empirical Approaches and Evidence (National Bureau of Economic Research, August 2013), pp. 14–19, www.nber.org/papers/w19377.

[17] Michael G. Finn, Stay Rates of Foreign Doctorate Recipients From U.S. Universities, 2009 (Oak Ridge Institute for Science and Education, January 2012), http://tinyurl.com/pelzz7l.

[18] The Optional Practical Training program allows some student visa holders to stay in the United States for up to an additional year after graduation, if they are working in their field of study, and students in STEM fields can apply to stay longer.

[19] See Lesleyanne Hawthorne, Competing for Skills: Migration Policies and Trends in New Zealand and Australia (New Zealand Department of Labour, 2011), pp. 108–114, http://tinyurl.com/kod79vz.

[20] However, to the extent that a patent-protected good faces competition from products that are similar to it, but do not apply the technology in question, the patent owner's ability to raise prices and restrict output through that monopoly will be diminished.

[21] Of particular relevance is ongoing research that explores how the importance of patents varies across industries and how strategies for protecting intellectual property can vary significantly across firms; see Bronwyn Hall and others, "The Choice Between Formal and Informal Intellectual Property: A Review," Journal of Economic Literature, vol. 52, no. 2 (June 2014), pp. 375–423, http://dx.doi.org/10.1257/jel.52.2.375.

[22] Several factors that are unrelated to the patent system's performance in supporting innovation could have contributed to the apparent discrepancy in Figure 3-1. For example, innovation may have taken the form of improvements to capital rather than TFP. Also, innovation in some fields may have experienced diminishing economic returns over time. Research has found that since about 1980, increases in computing power (and associated decreases in computing costs) as measured through technical performance criteria have outstripped economic measures of computing productivity (which are based on users' valuation of computing power relative to other features of computer systems); see William D. Nordhaus, "Two Centuries of Productivity Growth in Computing," Journal of Economic History, vol. 67, no. 1 (March 2007), pp. 151–154, http://tinyurl.com/k454h9v (PDF, 236 KB). Finally, the relationship between patenting and TFP growth could have weakened over time if innovation increasingly took the form of goods and services (such as pharmaceuticals) whose returns may generally not be well captured by economic indicators such as GDP. However, those effects do not appear to be important here, for two reasons. First, the average annual growth in capital services actually fell between the two periods in Figure 3-1. Second, much of the increase in patenting activity since the early 1990s can be attributed to software-related patents, including business-method patents. See Government Accountability Office, Intellectual Property: Assessing Factors That Affect Patent Infringement Litigation Could Help Improve Patent Quality, GAO-13-465 (August 2013), Figure 1, p. 12, www.gao.gov/products/GAO-13-465.

[23] See Stephen A. Merrill, Richard C. Levin, and Mark B. Myers, eds., A Patent System for the 21st Century (National Research Council of the National Academies, 2004), www.nap.edu/catalog.php?record_id=10976; Federal Trade Commission, To Promote Innovation: The Proper Balance of Competition and Patent Law and Policy (October 2003), http://go.usa.gov/ETjH (PDF, 2.3 MB); and Federal Trade Commission, The Evolving IP Marketplace: Aligning Patent Notice and Remedies with Competition (March 2011), http://go.usa.gov/ETD4 (PDF, 2.6 MB).

[24] See, for example, Adam B. Jaffe and Josh Lerner, Innovation and Its Discontents: How Our Broken Patent System Is Endangering Innovation and Progress, and What to Do About It (Princeton University Press, 2004); and Michael Heller, The Gridlock Economy: How Too Much Ownership Wrecks Markets, Stops Innovation, and Costs Lives (Basic Books, 2008).

[25] On the usefulness of patents as a tool for small firms to raise capital, see Stuart Graham and others, "High Technology Entrepreneurs and the Patent System: Results of the 2008 Berkeley Patent Survey," Berkeley Technology Law Journal, vol. 24, no. 4 (2009), pp. 1248–1318, http://btlj.org/category/journal/volume-24-2009/ and Michael J. Meurer, "Inventors, Entrepreneurs, and Intellectual Property Law," Houston Law Review, vol. 45, no. 4 (2008–2009), pp. 2001–2037, www.houstonlawreview.org/2009/01/. Data on the pendency of patent applications comes from United States Patent and Trademark Office, Performance and Accountability Report, Fiscal Year 2013, and Annual Report, Fiscal Year 1993, www.uspto.gov/about/stratplan/ar/.

[26] See Government Accountability Office, Intellectual Property: Assessing Factors That Affect Patent Infringement Litigation Could Help Improve Patent Quality, GAO-13-465 (August 2013), pp. 14–16, www.gao.gov/products/GAO-13-465. For a defense of the quality of software patents, see Stuart Graham and Saurabh Vishnubhakat, "Of Smart Phone Wars and Software Patents," Journal of Economic Perspectives, vol. 27, no. 1 (Winter 2013), pp. 67–86, http://dx.doi.org/10.1257/jep.27.1.67.

[27] Because the average PAE lawsuit has, historically, named more defendants than a patent infringement lawsuit brought by another type of plaintiff, litigation initiated by PAEs accounted for about one-half of the increase in the total number of patent infringement defendants from 2007 to 2011. See Government Accountability Office, Intellectual Property: Assessing Factors That Affect Patent Infringement Litigation Could Help Improve Patent Quality, GAO-13-465 (August 2013), p. 18, www.gao.gov/products/GAO-13-465.

[28] See James E. Bessen and Michael J. Meurer, The Direct Costs From NPE Disputes, Law and Economics Research Paper 12-34 (Boston University School of Law, revised July 2013), http://tinyurl.com/loqcl5f; Michael Risch, "Patent Troll Myths," Seton Hall Law Review (2012), http://works.bepress.com/michael_risch/16; and Colleen Chien, Patent Assertion and Startup Innovation (New America Foundation, September 2013), http://tinyurl.com/ob7k83j.

[29] See "Inside Views: Interview With Chief Judge Paul R. Michel on U.S. Patent Reform," Intellectual Property Watch (July 7, 2011), http://tinyurl.com/ndqvc9a.

[30] For example, the Federal Trade Commission is currently collecting information about PAEs to determine the impact that their activity has on competition in the market for patented products. See Federal Trade Commission, "Agency Information Collection Activities; Submission for OMB Review; Comment Request," Federal Register, vol. 79, no. 96 (May 19, 2014), pp. 28715–28729, http://go.usa.gov/7qKG (PDF, 241 KB).

[31] For roughly a decade beginning in March 2000, the USPTO required that business method patents be reviewed by two examiners. Although this resulted in a higher rejection rate (compared with other groups of patent applications for which there are a large number of submissions to the USPTO), it also considerably lengthened the examination time. See Mark A. Lemley, Fixing the Patent Office, Working Paper 18081 (National Bureau of Economic Research, May 2012), pp. 7–8, www.nber.org/papers/w18081.

[32] Research has found that roughly 28 percent of the patent applications from 1993 to 1998 were actually resubmissions of rejected patent applications; see Cecil D. Quillen Jr. and Ogden H. Webster, "Continuing Patent Applications and Performance of the U.S. Patent Office," Federal Circuit Bar Journal, vol. 11, no. 1 (August 2001), pp. 1–21; and United Kingdom Intellectual Property Office and United States Patent and Trademark Office, Patent Backlogs, Inventories and Pendency: An International Framework, Working Draft (June 26, 2013), pp. 68–69, www.ipo.gov.uk/ipresearch-uspatlog-201306.pdf (4 MB).

[33] A number of the proposals discussed in this section—in particular, those that would enable the courts to offer (limited) legal protections for end users and to implement fee-shifting—are among the provisions of the Innovation Act (H.R. 3309), which passed the House of Representatives on December 5, 2013.

[34] Another measure intended to reduce the cost of patent infringement litigation, which is incorporated in several current legislative proposals to modify the patent system, is to extend to software patents the AIA's Covered Business Method program. That would allow the validity of some software patents to be determined by administrative judges at the USPTO rather than by judges (and juries) in federal court; as a result, the cost of patent litigation in such cases should fall.

[35] See John R. Thomas, Tailoring the Patent System for Specific Industries, Report for Congress R43264 (Congressional Research Service, October 2013).

[36] For current USPTO efforts to improve the patent system, see U.S. Patent and Trademark Office, "Initiatives and Events" (last modified October 21, 2014), www.uspto.gov/patents /init_events/. Recent Supreme Court decisions have granted lower courts greater discretion to award attorney fees to prevailing parties in patent infringement lawsuits (Octane Fitness LLC. v. Icon Health and Fitness and Highmark, Inc., v. Allcare Health Management System); made it more difficult to assert a claim of patent infringement (Limelight Networks, Inc., v. Akamai Technologies, Inc.); heightened the requirement that a patent describe an invention in a clear and distinct way (Nautilus v. Biosig Instruments); and reaffirmed the ineligibility for patent protection of abstract ideas (Alice Corporation Pty., Ltd., v. CLS Bank International et al.).

[37] See Michael S. Barr, "The Financial Crisis and the Path of Reform," Yale Journal on Regulation, vol. 29, no. 1 (2012), pp. 91–119, http://tinyurl.com/oo8njtt.

[38] Testimony of Douglas W. Elmendorf, Director, Congressional Budget Office, before the Senate Committee on the Budget, Policies for Increasing Economic Growth and Employment in 2012 and 2013 (November 15, 2011), www.cbo.gov/publication/42717.

[39] Luke A. Stewart, The Impact of Regulation on Innovation in the United States: A Cross-Industry Literature Review (Information Technology and Innovation Foundation, June 2010), http://tinyurl.com/oo5qsf2 (PDF, 340 KB).

[40] The earlier deregulation of railroad rates also helped firms meet the sulfur reduction goals, as competition in rate setting sufficiently reduced the cost of shipping low-sulfur coal from the West to make it more viable. See Richard Schmalensee and Robert N. Stavins, The SO2 Allowance Trading System: The Ironic History of a Grand Policy Experiment, RFF Discussion Paper 12-44 (Resources for the Future, August 2012), pp. 8–10, http://tinyurl.com/pcxesz9. In this instance, deregulation in one area helped make market-oriented regulation in a different area successful.

[41] Testimony of Patrick A. McLaughlin, Senior Research Fellow, George Mason University, before the Subcommittee on Regulatory Reform, Commercial, and Antitrust Law of the House Committee on the Judiciary (February 11, 2014), http://tinyurl.com/n99d6om.

[42] See Richard Schmalensee and Robert N. Stavins, The SO2 Allowance Trading System: The Ironic History of a Grand Policy Experiment, RFF Discussion Paper 12-44 (Resources for the Future, August 2012), pp. 8–11, http://tinyurl.com/pcxesz9.

[43] Cary Coglianese, "Moving Forward With Regulatory Lookback," Yale Journal on Regulation, vol. 30, no. 57 (2013), pp. 57–66, http://tinyurl.com/pgld656.

[44] For other examples of federal efforts to mitigate state court tort litigation, see James M. Anderson and others, Autonomous Vehicle Technology: A Guide for Policymakers (RAND Corporation, 2014), pp. 131–132, www.rand.org/pubs/research_reports/RR443-1.html.

[45] See National Highway Traffic Safety Administration, "Early Estimate of Motor Vehicle Traffic Fatalities in 2013," DOT HS 812 024 (May 2014), http://go.usa.gov/wn7w.

[46] See John Villasenor, Products Liability and Driverless Cars: Issues and Guiding Principles for Legislation (Brookings Institution, April 2014), http://tinyurl.com/k6921d3.

[47] See Kathleen Ann Ruane, Broadband Deployment: Legal Issues for the Siting of Wireless Communications Facilities and Amendments to the Pole Attachment Rule, Report for Congress RS20783 (Congressional Research Service, April 2013).

In: Federal Influence on Innovation
Editor: Rachael A. Young
ISBN: 978-1-63463-627-8

Chapter 2

RESEARCH TAX CREDIT: CURRENT LAW AND POLICY ISSUES FOR THE 113TH CONGRESS*

Gary Guenther

SUMMARY

Technological innovation is a primary engine of long-term economic growth, and research and development (R&D) serves as the lifeblood of innovation. The federal government encourages businesses to invest more in R&D than they otherwise would in several ways, including a tax credit for increases in spending on qualified research above a base amount.

This report describes the current status of the credit, summarizes its legislative history, discusses policy issues it raises, and describes legislation to modify and extend it.

The research credit (also known as the research and experimentation (R&E) tax credit) has never been permanent. It expired at the end of 2011 and was retroactively extended by the American Taxpayer Relief Act of 2012 (P.L. 112-240) through the end of 2013. Since its enactment in mid-1981, the credit has been extended 15 times and significantly modified 5 times.

* This is an edited, reformatted and augmented version of a Congressional Research Service publication RL31181, prepared for Members and Committees of Congress, dated October 3, 2014.

While the credit is usually assumed to be a single credit, it actually consists of four discrete credits: (1) a regular credit, (2) an alternative simplified credit (ASC), (3) a basic research credit, and (4) an energy research credit. A taxpayer may claim one of the first two and each of the other two, provided it meets the requirements for each.

In essence, the research credit attempts to boost business investment in basic and applied research by reducing the after-tax cost of undertaking qualified research above a base amount, which in theory approximates the amount a company would invest in R&D in the absence of the credit. As a result, the credit's effectiveness hinges on the sensitivity of the demand for this research to decreases in its cost. It is unclear from available studies how sensitive that demand actually is.

While most analysts and lawmakers endorse the use of tax incentives to generate increases in business R&D investment, some have some reservations about the current credit. Critics contend that it is not as effective as it could or should be because of certain problems with its design. These include a lack of permanence, uneven and inadequate incentive effects, non-refundability, and an ambiguous definition of qualified research.

The House has passed two bills (H.R. 4438, the American Research and Competitiveness Act of 2014 on May 9 and H.R. 4, the Jobs for America Act on September 18) that would simplify and permanently extend the research tax credit. Under each measure, the credit would be equal to the sum of 20% of a taxpayer's qualified research expenditures (QREs) in the current tax year above 50% of average annual QREs in the previous three tax years, 20% of its basic research payments in the current tax year above 50% of average annual basic research payments in the three previous tax years, and 20% of the amounts paid or incurred by the taxpayer in the current tax year for qualified energy research conducted by an energy research consortium. The bills' estimated 10-year revenue cost is $155.5 billion.

On April 3, the Senate Finance Committee marked up S. 2260, the Expiring Provisions Improvement, Reform, and Efficiency Act. Among other things, the bill would extend the existing research credit through the end of 2015. It would also change its design by allowing small, young companies that are unable to use some or all of any credit they could claim for the current tax year because of insufficient tax liability to apply up to $250,000 of their credit against their federal payroll tax liabilities. This option would be available only to companies that have been in business fewer than five years and whose annual gross receipts are less than $5 million.

INTRODUCTION

Economists have gained notoriety for their differences of opinion on a variety of policy issues. Notable examples include the long-term economic effects of large, permanent tax cuts; the impact of illegal immigration on domestic wages; and the best way to achieve price stability, full employment, and greater income equality. But on the issues of the impact of technological innovation on economic growth in the long run and the proper role of government in the development of new technologies, there is relatively little dissent among practitioners of what is known as the dismal science.

In general, economists agree that technological innovation has accounted for a major share of long-term growth in real per-capita income in the United States.[1] It is useful to clarify what economists mean by technological innovation, since such a complex concept can mean different things to different professions. Economists who study the forces driving economic growth tend to see innovation as a convoluted and uncertain process that embraces the acquisition of new scientific and technical knowledge and its application to the development of new goods and services or methods of production through research and experimentation. Learning-by-doing and learning-by-using often play crucial roles in this process.

In economies dominated by competitive markets, technological innovation is driven mainly by the efforts of competing firms to gain, sustain, or reinforce competitive advantages by being the first to introduce or use new or improved products or services; more efficient production processes; or more effective strategies for management, marketing and promotion, and customer service and support.

Most economists would also agree that private R&D investment is likely to be less than the amounts that would be warranted by its economic benefits. The reason for this is thought to lie in the nature of these benefits. Firms generally cannot capture all the returns to their R&D investments, even in the presence of patents, trademarks, and other instruments of intellectual property protection, and their strict enforcement. Numerous studies have found that the average social returns to private R&D investments greatly exceeded the average private returns.[2] This finding held true whether a firm invested in research projects narrowly focused on its existing lines of business, or in research projects aimed at extending the boundaries of knowledge in particular scientific disciplines in ways that had no obvious or immediate commercial applications.

Economists refer to an excess of social returns over private returns as the spillover effects or external benefits of R&D. There are several channels through which the returns from innovation may escape full capture by the innovating firms and spill over to society at large. The most common ones are reverse engineering by competing firms, migration of research scientists and engineers from one firm to another, and the availability of new or improved goods and services at prices below what most consumers would be willing to pay.[3]

When seen through the lens of standard economic theory, the external benefits of technological innovation take on the appearance of a market failure. They signal that too few resources are being allocated to the activities leading to the discovery and commercial development of new technical knowledge and know-how. To remedy this failure, most economists recommend the adoption of public policies aimed at boosting or supplementing private investment in R&D.

The federal government supports R&D in a variety of ways. Direct support comes mainly in the form of research performed by federal agencies and federal grants for basic and applied research and development intended to support concrete policy goals, such as protecting the natural environment, exploring outer space, advancing the treatment of deadly diseases, and strengthening the national defense. Indirect support is more diffuse. The chief sources are federal funding of higher education in engineering and the natural sciences, legal protection of intellectual property rights, special allowances under antitrust law for joint research ventures, and tax incentives for business R&D investment.

Federal tax law offers two such incentives: (1) an unlimited expensing allowance for qualified research spending under Section 174 of the Internal Revenue Code (IRC), and (2) a nonrefundable tax credit for qualified research spending above a base amount under IRC Section 41—known as the research and experimentation (R&E) tax credit, the research tax credit, the R&D tax credit, or the credit for increasing research activities. The deduction has been a permanent provision of the IRC since it was first enacted in 1954. Its main advantages are that the deduction simplifies tax accounting for R&D expenditures and encourages business R&D investment by taxing the returns to such investment at a marginal effective rate of zero. A similar policy objective lies behind the research tax credit, which has been a temporary provision of the IRC since it went into effect in July 1981. The credit is intended to stimulate more business R&D investment than otherwise would occur by lowering the after-tax cost of qualified research.[4] But unlike the deduction, it complicates tax compliance for firms claiming the credit. In FY2012, the combined budgetary cost of these incentives totaled $11.1 billion, or about 8% of the estimated $140.9 billion spent by federal agencies on defense and non-defense R&D that year.[5]

This report examines the current status of the R&E tax credit, describes its legislative history, discusses some important policy issues raised by it, and identifies legislative proposals in the 113th Congress to extend or otherwise modify the credit.

DESIGN OF THE R&E TAX CREDIT

Although many think of the research tax credit as a single unified credit, it has four discrete components: a regular research credit, an alternative simplified credit (ASC), a basic research credit, and a credit for energy research.[6] Each is non-refundable. In any tax year, taxpayers may claim no more than the basic and energy research credits, plus either the regular credit or the ASC. To prevent taxpayers from benefiting twice from the same expenditures, any research tax credit claimed must be subtracted from deductible research expenses. The four components of the research tax credit expired at the end of 2013.

Qualified Research Expenditures

Ultimately, claims for the regular credit and the ASC rest on the definition of qualified research expenditures (QREs). There are two aspects to this definition.

Nature of Qualified Research

One aspect deals with the nature of qualified research itself. Under Section 41(d) of the federal tax code, research must satisfy four criteria in order to qualify for the regular credit and the ASC. First, the research must involve activities that qualify for the deduction under Section 174: namely, the activities must be "experimental" in the laboratory sense and aimed at the development of a new or improved product or process. Second, the research must seek to discover information that is "technological in nature." Third, it should strive to gain new technical knowledge that is useful in the development of a new or improved "business component," which is defined as a product, process, computer software technique, formula, or invention to be sold, leased, licensed, or used by the firm performing the research. And fourth, the research must entail a process of experimentation aimed at the development of a product or process with "a new or improved function, performance or reliability or quality."

According to Section 41(d)(3), research satisfies the four criteria if it is intended to develop a new or improved function for a business component, or to improve the performance, reliability, or quality of a business component. By contrast, research fails to meet these criteria if its main purpose is to modify a business component according to "style, taste, (and) cosmetic or seasonal design factors."

Businesses, the courts, and the IRS have clashed repeatedly over the interpretation of the four criteria. Although the IRS issued final regulations clarifying the definition of qualified research in December 2003 (T.D. 9104), disputes between businesses and the IRS over what activities qualify for the credit have continued.[7]

Section 41(d)(4) identifies the activities for which the credit may not be claimed. Specifically, the credit does not apply to:

- research conducted after the start of commercial production of a "business component";
- research done to adapt an existing business component to a specific customer's needs or requirements;
- research related to the duplication of an existing business component;
- surveys and studies related to data collection, market research, production efficiency, quality control, and managerial techniques;
- research to develop computer software for a firm's internal use (except as allowed in any regulations issued by the IRS);
- research conducted outside the United States, Puerto Rico, or any other U.S. possession;
- research in the social sciences, arts, or humanities; and
- research funded by another entity.

Expenses Eligible for the Credit

The second aspect of the definition of QREs concerns the expenses to which the credit applies. Under Section 41(b)(1), qualified expenses relate to both in-house research and contract research. In the case of in-house research, the regular credit and ASC apply to the wages and salaries of employees and supervisors engaged in qualified research, as well as the cost of materials, supplies, and leased computer time used in this research. In the case of contract research, the credits apply to the full amount paid for qualified research conducted by certain small firms, colleges and universities, and federal laboratories; 75% of payments for qualified research performed by certain research consortia; and 65% of payments for qualified research performed by certain other non-profit entities dedicated to scientific research.

As a result, the credits do not cover all the expenses a company incurs in conducting qualified research. Specifically, outlays for depreciable durable assets used in qualified research (such as buildings and equipment), overhead expenses (e.g., heating, electricity, rents, leasing fees, insurance, and property taxes), and the fringe benefits of research personnel are excluded. The exclusion of these expenses has implications for the incentive effect of the credit (more on this later). According to some estimates, the excluded expenses account for 27% to 50% of business R&D spending.[8]

Regular Research Credit

The regular research tax credit has been extended 15 times and significantly modified 5 times. Under IRC Section 41(a)(1), it is equal to 20% of a firm's QREs beyond a base amount. Such an incremental design is intended to encourage firms to spend more on R&D than they otherwise would by lowering the after-tax cost to business taxpayers of investing in qualified research above some normal amount by as much as 20%.[9] Given that business R&D investment appears sensitive to its cost, a decline in the after-tax cost of R&D can be expected to spur a rise in business R&D investment, all other things being equal.[10]

The base amount for the regular credit is designed to approximate the amount a firm would spend on qualified research in the absence of the credit. As such, the base amount can be viewed as a firm's normal or preferred level of R&D investment. Two rules govern the calculation of the base amount under IRC Section 41(c). First, it must be equal to 50% or more of a firm's QREs in the current tax year—a rule that some refer to as the 50-percent rule.[11] Second, the base amount depends on whether the business taxpayer is considered an established firm or a start-up firm. Established firms are defined as firms with gross receipts and QREs in at least three of the tax years from 1984 through 1988. Start-up firms, by contrast, are defined as firms whose first tax year with both gross receipts and QREs occurred after 1983, or firms that had fewer than three tax years from 1984 to 1988 with both gross receipts and QREs.[12] The base amount for all firms, established or start-up, is the product of a fixed-base percentage and average annual gross receipts in the previous four tax years. An established firm's fixed-base percentage is the ratio of its total QREs to total gross receipts in 1984 to 1988, capped at 16%. By contrast, a start-up firm's fixed-base percentage is set at 3% during the firm's first five tax years with spending on qualified research and gross receipts. Thereafter, the percentage gradually adjusts to reflect a firm's actual experience, so that by its 11th tax year, the percentage equals the firm's total QREs relative to its total receipts in the 5th through 10th tax years.

In general, the lower a firm's fixed-base percentage, the better its chances of claiming the regular credit. Furthermore, a firm can expect to benefit from the regular credit if its ratio of QREs in the current tax year to average annual gross receipts in the previous four tax years is greater than its fixed-base percentage. (See Table 1 for a calculation of the regular credit for a hypothetical established firm and Table 2 for a calculation of the regular credit for a hypothetical start-up firm.)

Table 1. Sample Calculations of the Regular and Alternative Simplified Research Tax Credits in 2014 for an Established Firm ($ millions)

Year	Gross Receipts	Qualified Research Expenses
1984	100	5
1985	150	8
1986	250	12
1987	400	15
1988	450	16
1989	400	18
1990	450	18
2007	835	45
2008	915	50
2009	1,005	53
2010	1,215	60
2011	1,465	70
2012	1,650	85
2013	1,825	95
2014	1,900	100

Source: Congressional Research Service.

Calculation: Regular Research Tax Credit

Compute the fixed-base percentage:

1) Sum the qualified research expenses for 1984 to 1988: $56 million.
2) Sum the gross receipts for 1984 to 1988: $1,350 million.
3) Divide the total qualified research expenses by the total gross receipts to determine the fixed-base percentage: 4.0%.

Compute the base amount for 2014:

1) Calculate the average annual gross receipts for the four previous years (2010-2013): $1,539 million.
2) Multiply this average by the fixed-base percentage to determine the base amount: $62 million.

Compute the regular tax credit for 2014:

1) Reduce the $100 million in qualified research expenses for 2014 by the greater of the base amount ($62 million) or 50% of the qualified research expenses for 2014 ($50 million): $38 million.
2) Multiply this amount by 20% to determine the regular R&E tax credit for 2014: *$7.60 million.*

Calculation: Alternative Simplified Research Credit

1) Calculate the average qualified research expenditures in the three previous years (2011- 2013): $83 million.
2) Divide this amount by 2: $41.5 million.
3) Subtract this amount from qualified research expenditures in 2014: $58.5 million.
4) Multiply this amount by 0.14 to determine the alternative simplified research credit for 2014: $8.2 million.

Calculation: Regular Research Tax Credit

Compute the fixed-base percentage:

1) According to current law, a start-up firm's fixed-base percentage is fixed at 3% for each of the first five years after 1993 when it has both gross receipts and qualified research expenses; it then adjusts

according to a formula over the next six years to reflect the firm's actual research intensity. Thus, the fixed-base percentages are 3% for 2006 through 2010, 7.4% in 2011, 8.9% in 2012, 12.0% in 2013, and 14.9% in 2014.

Table 2. Sample Calculations of the Regular and Alternative Simplified Research Tax Credits in 2014 for a Start-up Firm ($ millions)

Year	Gross Receipts	Qualified Research Expenses
2006	30	35
2007	42	40
2008	55	45
2009	60	55
2010	210	65
2011	305	73
2012	400	82
2013	475	90
2014	600	105

Source: Congressional Research Service.

Compute the base amount for 2014:

1) Calculate the average annual receipts for the four previous years (2010-2013): $347.5 million.
2) Multiply this amount by the fixed-base percentage (14.9%) to determine the base amount: $52 million.

Compute the regular tax credit:

1) Reduce qualified research expenses for 2014 ($105 million) by the greater of the base.
2) Amount ($52 million) or 50% of the qualified research expenses for 2014 ($52.5 million): $52.5 million.
3) Multiply this amount by 20% to determine the regular R&E tax credit for 2014: *$10.5 million.*

Calculation: Alternative Simplified Research Credit

1) Calculate the average qualified research expenditures for the three previous years (2011- 2013): $82 million.
2) Divide that amount by 2: $41 million.
3) Subtract that amount from qualified research expenditures in 2014: $64 million.
4) Multiply this amount by 0.14 to determine the alternative simplified research credit for 2014: *$9.0 million.*

Alternative Simplified Credit

The most recent addition to the array of research tax credits provided by Section 41 is the alternative simplified credit (ASC). It was established by the Health Care and Tax Relief Act of 2006 (P.L. 109-432). Under Section 41(c)(5), a business taxpayer may claim the ASC in lieu of the regular credit. The ASC is equal to 14% of a taxpayer's QREs in the current tax year above 50% of its average QREs during the three previous tax years. If a taxpayer has no QREs in any of those years, then the credit is equal to 6% of its QREs in the current tax year. A decision to elect the ASC remains in effect until a taxpayer gains the consent of the IRS to claim the regular research credit. (See Table 1 for a hypothetical calculation of the ASC for an established firm and Table 2 for a similar calculation of the ASC for a startup firm.)

Taxpayers with one or more of the following conditions may benefit more from the ASC than the regular credit:

- a relatively large base amount under the regular credit;
- incomplete records for determining the base period as a start-up firm;
- substantial growth in gross receipts in recent years; and
- a complicated history of mergers, re-organizations, acquisitions, and dispositions.

Alternative Incremental Research Credit

Firms investing in qualified research that could not claim the regular credit had the option of claiming the alternative incremental R&E tax credit (or AIRC), under IRC Section 41(c)(4), for tax years from 1996 to 2008. The Emergency Economic Stabilization Act of 2008 (P.L. 110-343) repealed the AIRC for the 2009 tax year, and Congress has not reinstated it. When a firm elected the AIRC for a particular tax year, it had to continue to do so, unless the firm received permission from the IRS to switch to the regular research credit. There was some concern that such a rule deterred firms from claiming the AIRC, even though they might have been better off doing so.

The definition of QREs for the AIRC was the same as the definition of QREs for the regular credit. But that was where the similarity between the two credits ended. While the regular credit is equal to 20% of QREs in excess of a base amount, the AIRC, in the final year it could be taken, was equal to 3% of a firm's QREs above 1% but less than 1.5% of its average annual gross receipts in the previous four tax years, plus 4% of its QREs above 1.5% but less than 2.0% of its average annual gross receipts in the previous four tax years, plus 5% of its QREs greater than 2.0% of its average annual gross receipts in the previous four tax years.

In general, firms benefited from the AIRC if their QREs in the current tax year exceeded 1% of their average annual gross receipts during the past four tax years. Moreover, the AIRC was probably of greater benefit than the regular credit was to business taxpayers with relatively high fixed-base percentages, or whose research spending was declining, or whose sales were growing much faster than their research spending. (See Table 1 for a calculation of the AIRC for a hypothetical established firm, and Table 2 for a calculation of the AIRC for a hypothetical startup firm.)

University Basic Research Credit

Firms that enter into contracts with certain non-profit organizations to perform basic research may be able to claim a separate research credit for some of their expenditures for this purpose under IRC Section 41(e). A primary aim of the credit is to foster collaborative research involving U.S. firms and colleges and universities. The credit is equal to 20% of total payments for qualified basic research above a base amount, which is known as the "qualified organization base period amount." This amount has little in common with the base amount for the regular R&E tax credit, except that both amounts seem intended to approximate what firms would spend on qualified research in the absence of the credits.[13]

For the purpose of the credit, basic research is defined as "any original investigation for the advancement of scientific knowledge not having a specific commercial objective."

The credit does not apply to qualified basic research done outside the United States, or to basic research in the social sciences, arts, or the humanities.

In addition, the basic research credit applies only to payments for qualified basic research performed under a written contract by the following organizations: educational institutions, nonprofit scientific research organizations (excluding private foundations), and certain grant-giving organizations.

Firms conducting their own basic research may not claim the credit for their expenditures for this purpose, but the spending may be included in their QREs for the regular credit or ASC. In addition, basic research payments eligible for the credit that fall below the base amount are treated as contract research expenses and may be included in the QREs for those credits as well.

Energy Research Credit

Under IRC Section 41(a)(3), taxpayers may claim a tax credit equal to 20% of payments to certain entities for energy research. To qualify for the credit, the payments must be made to a nonprofit organization exempt from taxation under IRC Section 501(a) and "organized and operated primarily to conduct energy research in the public interest." In addition, at least five different entities must contribute funds to the organization for energy research in a calendar year; none of these entities may account for more than half of total payments to the organization for qualified research.

The credit also applies to the full amount (i.e., 100%) of payments to colleges and universities, federal laboratories, and certain small firms for energy research performed under contract. In the case of small firms performing this research, a business taxpayer may claim a credit for the full amount of payments with two limitations. First, the taxpayer cannot own 50% or more of the stock of the small firm performing the research (if the firm is a corporation), or hold 50% or more of the small firm's capital and profits (if the firm is a non-corporate entity such as a partnership). Second, the firm performing the research must have an average of 500 or fewer employees in one of the two previous calendar years.

Because the credit is flat rather than incremental, it is more generous than the other four components of the research tax credit.

Option to Claim a Refundable Research Tax Credit in Lieu of Bonus Depreciation in 2008 and 2009

As a result of the Economic Stimulus Act of 2008 (P.L. 110-185), corporate and non-corporate firms could claim an additional first-year depreciation deduction equal to 50% of the cost of qualified property placed in service between March 31, 2008, and December 31, 2008. The deduction was known as the 50% bonus depreciation allowance. A provision of the Housing and Economic Recovery Act of 2008 (P.L. 110-289) gave corporations only the option of claiming a limited refundable tax credit for unused research and alternative minimum tax (AMT) credits stemming from tax years before 2006, in lieu of any bonus depreciation allowance they could claim for qualified property acquired after March 31, 2008. The credit was capped at $30 million for a single corporation and was set to expire at the end of 2008.

The American Recovery and Reinvestment Act of 2009 (ARRA, P.L. 111-5) extended both the first-year 50-percent bonus depreciation allowance and the option to claim a refundable research and AMT credit through 2009.

Under the Tax Relief, Unemployment Compensation Reauthorization, and Job Creation Act of 2010 (P.L. 111-312), the option to monetize unused AMT credits from tax years before 2006 in lieu of claiming a bonus depreciation allowance was extended so that it applied to qualified property acquired after March 31, 2008, and before January 1, 2013. The extension did not apply to unused research credits from the same tax years. With the passage of the American Taxpayer Relief Act of 2012 (P.L. 112-240), the option was extended through 2013 for qualified property acquired and placed in service that year.

LEGISLATIVE HISTORY OF THE RESEARCH TAX CREDIT

The research tax credit entered the tax code as a temporary provision through the Economic Recovery Tax Act of 1981 (P.L. 97-34). In adopting the credit, the 97th Congress was hoping to stem a decline in spending on R&D by the private sector as a share of U.S. gross domestic product that commenced in the late 1960s. Around the time the credit was enacted, more than a few analysts thought the decline was a primary cause of both a slowdown in U.S. productivity growth and an unexpected loss of competitiveness by a variety of U.S. industries in the 1970s. A majority in Congress concluded that a "substantial tax credit for incremental research and experimental expenditures was needed to overcome the reluctance of many ongoing companies to bear the significant costs of staffing and supplies, and certain equipment expenses such as computer charges, which must be incurred to initiate or expand research programs in a trade or business."[14]

The initial credit was equal to 25% of qualified research spending above a base amount, which was equal to average spending on such research in the three previous tax years, or 50% of current-year spending, whichever was greater. It is not clear from the historical record why Congress chose a statutory rate of 25%. But there is no evidence that the rate was chosen on the basis of a rigorous assessment of the gap between the private and social returns to R&D investment, or the sensitivity of R&D expenditures to declines in their after-tax cost. Any taxpayer that claimed the credit and could not apply the entire amount against its current-year federal income tax liability was allowed to carry the unused portion back as many as three tax years, or forward as many as 15 tax years. The credit was to remain in effect from July 1, 1981, to December 31, 1985.

Congress made the first significant changes in the original research tax credit with the passage of the Tax Reform Act of 1986 (TRA86, P.L. 99-514). Among the many significant changes it made in the federal tax code, the act extended the credit through December 31, 1988, and folded it into the general business credit under IRC Section 38, thereby subjecting it to a yearly cap. In addition, the act lowered the credit's statutory rate to 20%, modified the definition of QREs so that the credit applied to research intended to produce new technical knowledge deemed useful in the commercial development of new products and processes, and created a separate 20% incremental tax credit for payments to universities and certain other nonprofit organizations for the conduct of basic research according to a written contract. The reduction in the credit's rate seemingly was not based on an analysis of the credit's effectiveness in the first five years. Rather, it seemed to stem from the overriding goals of TRA86, which were to lower income tax rates across the board, broaden the income tax base, and shrink the differences in tax burdens among major categories of business investments. Since 1954, firms investing in R&D had already been benefiting from the option to expense qualified R&D spending under the IRC Section 174 expensing allowance.[15]

The regular and basic research credits were further altered by the Technical and Miscellaneous Revenue Act of 1988 (P.L. 100-647). Specifically, the act extended the credits through December 31, 1989. It also curtailed the overall tax preference for private-sector R&D investment by requiring business taxpayers to reduce any deduction they claim for research spending under IRC Section 174 by half of the sum of any regular and basic research credits they claim. This new rule decreased the maximum effective rate of the regular research tax credit by a factor equal to 0.5 times a taxpayer's marginal income tax rate.[16]

Continuing disappointment with the design of the original credit among interested parties led to the enactment of several additional changes in the regular credit through the Omnibus Budget Reconciliation Act of 1989 (OBRA89, P.L. 101-239). Much of the disappointment was focused on the formula for determining the base amount of the credit. Critics rightly pointed out that under the formula, which was based on a three-year moving average of a firm's annual spending on qualified research, an increase in a firm's research spending in one year would raise its base amount in each of the following three years by an amount equal to one-third of that increase, possibly making it more difficult to claim the credit in those three years. Some argued that such a design would be less cost-effective in boosting business R&D investment than one in which a firm's base amount was completely independent of its current spending on qualified research.[17]

To address this concern, OBRA89 changed the formula for the base amount so that it was equal to the greater of 50% of a firm's current-year QREs, or the product of the firm's average annual gross receipts in the previous four tax years and a "fixed-base percentage." The act set this percentage equal to the ratio of a firm's total QREs to total gross receipts in the four tax years from 1984 to 1988, capped at 16%. OBRA89 also made the credit available on more favorable terms to start-up firms, which it defined as firms without gross receipts and QREs in three of the four years from 1984 to 1988; these firms were assigned a fixed-base percentage of 3%. In addition, the act effectively extended the credits to December 31, 1990 (by requiring that QREs incurred before January 1, 1991, be prorated), made it clear that firms could apply the regular credit to QREs related to current lines of business and possible future lines of business, and required firms claiming the regular and basic research credits to reduce any deduction they claim under IRC Section 174 by the entire amount of the credits.

In 1990 and 1991, Congress passed two bills that, among other things, temporarily extended the credits. The Omnibus Budget Reconciliation Act of 1990 (P.L. 101-508) extended the credits through December 31, 1991, and repealed the requirement that QREs made before January 1, 1991, be prorated. And the Tax Extension Act of 1991 (P.L. 102-227) pushed the expiration date for the credits forward to June 30, 1992. A major obstacle to longer extensions of the credits at the time lay in congressional budget rules requiring that the revenue cost of lengthy or permanent extensions be scored over 10 fiscal years and offset with tax increases or cuts in non-defense discretionary spending.

Although Congress passed two bills in 1992 that would have extended the credits beyond June 30 of that year, President George H. W. Bush vetoed both for reasons that had nothing to do with the credit provisions. As a result, the credits expired and remained unavailable from July 1, 1992, until the enactment of the Omnibus Budget Reconciliation Act of 1993 (OBRA93, P.L. 103-66) in August 1993. That act extended the credits retroactively from July 1, 1992, through June 30, 1995. It also modified the fixed-base percentage for start-up firms. Specifically, a firm lacking gross receipts in three of the years from 1984 to 1988 was assigned a percentage of 3% for the first five tax years after 1993 in which it reported QREs. Starting in the sixth year, the percentage was to adjust gradually so that by the 11th year the percentage would reflect its actual ratio of total QREs to total gross receipts in five of the previous six tax years.

Congressional inaction allowed the credits to expire again on June 30, 1995. They remained in abeyance until the enactment of the Small Business Job Protection Act of 1996 (P.L. 104-188) in August 1996. That act retroactively reinstated the credits from July 1, 1996, to May 31, 1997, leaving a one-year gap in the credit's coverage since its inception in mid-1981. The act also expanded the definition of a start-up firm to include any firm whose first tax year with both gross receipts and QREs was 1984 or later, added a three-tiered alternative incremental research credit (AIRC) with initial rates of 1.65%, 2.2%, and 2.75%, and made 75% of payments for qualified research performed under contract by nonprofit organizations "operated primarily to conduct scientific research" eligible for the regular credit and the AIRC.

The credits expired again in 1997, but they were extended retroactively from June 1, 1997, to June 30, 1998, by the Taxpayer Relief Act of 1997 (P.L. 105-34). A further extension of the credits to June 30, 1999, was included in the revenue portion of the Omnibus Consolidated and Emergency Supplemental Appropriations Act, 1998 (P.L. 105-277).

Under circumstances reminiscent of 1997, the credits expired in 1999. But the revenue portion of the Ticket to Work and Work Incentives Improvement Act of 1999 (P.L. 106-170) extended them from July 1, 1999, to June 30, 2004. It also increased the three rates of the AIRC to 2.65%, 3.2%, and 3.75% and expanded the definition of qualified research to include qualified research performed in Puerto Rico and the other territorial possessions of the United States.

On October 4, 2004, President George W. Bush signed into law the Working Families Tax Relief Act of 2004 (P.L. 108-311), which included a provision extending the research tax credit through December 31, 2005.

The Energy Policy Act of 2005 (P.L. 109-58) added a fourth component to the research tax credit by establishing a credit equal to 20% of payments for energy research performed under contract by qualified research consortia, colleges and universities, federal laboratories, and eligible small firms.

Under the Tax Relief and Health Care Act of 2006 (P.L. 109-432), the research tax credit was extended retroactively through the end of 2007. The act also raised the three rates for the AIRC to 3%, 4%, and 5%, and established yet another research tax credit, known as the alternative simplified credit (ASC). This fifth component of the credit was equal to 12% of QREs in excess of 50% of average QREs in the past three tax years; for business taxpayers with no QREs in any of the three preceding tax years, the credit was equal to 6% of QREs in the current tax year.

The Emergency Economic Stabilization Act of 2008 (P.L. 110-343) retroactively extended the research credit through 2009. It also raised the rate of the ASIC from 12% to 14% and repealed the AIRC for the 2009 tax year only.

Under the Housing and Economic Recovery Act of 2008 (P.L. 110-289), corporations gained the option for the 2008 tax year only of claiming a limited refundable credit for unused research and AMT credits from tax years before 2006, in lieu of any bonus depreciation allowance they could claim for qualified assets placed in service between March 31, 2008, and December 31, 2008.

The American Recovery and Reinvestment Act of 2009 (P.L. 111-5) extended that option through 2009, along with the bonus depreciation allowance that applied in 2008.

As a result of the Tax Relief, Unemployment Compensation Reauthorization, and Job Creation Act of 2010, (P.L. 111-312), the research credit was extended through 2011.

After a one-year lapse, Congress retroactively extended the credit through 2013 and made some minor changes in the rules governing the allocation of research credits among members of controlled groups of companies and the use of the credit by the parties to business acquisitions by passing the American Taxpayer Relief Act of 2012 (P.L. 112-240).

Going back to the mid-1990s, a cycle seems to emerge every time the credit is about to expire. The cycle commences when congressional and business supporters of the credit issue public statements calling for a permanent extension of the credit and denouncing what they see as the folly of repeated temporary extensions.[18] Then the President expresses his support for such an extension. In the next stage of the cycle, leaders in both houses of Congress enter into negotiations on tax legislation that may include a permanent extension of the credit. But in the end, Congress and the President can agree only on a limited extension of the credit, stymied by the difficulty of reconciling the revenue cost of a permanent extension with other budget priorities. The main elements of the cycle seemed to reappear in the round of negotiations between Congress and President Obama over legislation to avoid across-the-board rises in individual tax rates at the end of 2012.

EFFECTIVENESS OF THE RESEARCH TAX CREDIT

For analysts and lawmakers alike, probably the most important policy issue raised by the research tax credit is its effectiveness. Basically, there are two approaches to assessing this critical aspect of the credit.

Among economists, the preferred approach is to compare the social benefits from any added qualified research induced by the credit with the social costs of that research. Such an undertaking involves comparing the returns to society of the additional qualified research spending associated with the credit to the opportunity costs to society of those outlays. The social cost of the credit can be thought of as the net loss of tax revenue because of the credit, together with the public and private costs of administering the credit. Unfortunately, this approach has little application in policymaking, largely because it is difficult to measure accurately the social returns to R&D.[19]

As a result, economists have relied on the next best approach: estimating the additional qualified research (if any) stimulated by the regular credit, and comparing the dollar value of that gain with the tax revenue lost because of the credit. Such an approach compares the direct benefits (i.e., added research investment) with the direct costs (revenue loss) of the regular credit. It presupposes that the social returns to the research far exceed the private returns, and that the optimal size of any tax subsidy for R&D can be estimated. If the ratio of benefits to cost is greater than one, then the credit can be seen as a more cost-effective way to boost research than direct research subsidies; if it is less than one, then funding the research directly would be more cost-effective.[20]

What do available studies say about the effectiveness of the regular research credit? Going back to the early days of the credit, economists who have evaluated its impact on business investment in qualified research have used a variety of methods to estimate the amount of additional research that can be attributed to the credit. These methods were reviewed in a 1995 study by economist Bronwyn Hall.[21] She found that the studies using data on the credit from 1981 to 1983 came up with lower estimates of the added research associated with one dollar of the credit than did the studies based on data from 1984 and after. Taking into consideration the strengths and weaknesses of the studies, Hall concluded that the credit led to a "dollar-for-dollar increase in reported R&D spending on the margin."[22] In other words, over the period (mid-to-late 1980s) covered by the later studies, she found that one dollar of the credit resulted in a company spending an additional dollar on qualified research, suggesting that the credit was as cost-effective as federal research grants. It is not clear if the credit remains as cost-effective, or if it has become more or less so.

STIMULATIVE EFFECT OF THE CREDIT

What does available evidence say about the stimulative effect of the credit? This effect refers to the percentage increase in research spending over a period that can be attributed to the credit. Answering the question entails applying what is known about the responsiveness of business spending on research to declines in its tax price to the reduction in the aggregate cost of research that could be due to the credit.

In theory, the credit stimulates increased investment in qualified research by lowering the after-tax cost of undertaking another dollar of research. In other words, the credit lowers the tax price of research, and firms respond to the reduced prices by spending more on it, all other things being equal. Early studies of the responsiveness of research spending to price decreases estimated that the tax price elasticity of demand for research was less than one, perhaps somewhere between - 0.2 to -0.5. This meant that a 10% reduction in the after-tax (or net) cost of research would lead to an increase in research spending of 2% to 5%. More recent studies, however, indicated that research spending was more sensitive to price changes. On the whole, they suggested that the tax price elasticity was closer to -1.0 in the short run, which meant that a 10% reduction in the cost of research would trigger a 10% rise in research spending. Several of the studies also concluded that research tax credits may have an even larger impact on research spending in the longer run, on the order of -2.0.

Still, the question of how responsive the demand for research is to tax price decreases remains open to further debate and analysis. Existing studies have data and methodological limitations that cast serious doubt on the validity of their estimates. In an assessment of the revenue effects of President Bush's FY2004 budget request, the Joint Committee on Taxation (JCT) noted that "the general consensus when assumptions are made with respect to research expenditures is that the price elasticity of research is less than -1.0 and may be less than -0.5."[23] Therefore, considering the findings of the studies that have been done, it seems reasonable to assume that the short-run price elasticity for qualified research lies between -0.4 and -0.8.

A measure of the reduction in the cost of qualified research that can be attributed to the credit is its average effective rate. The rate captures the percentage reduction in that cost that is due to the credit. It is derived by dividing the total amount of claims for the research credit in a tax year by some measure of total business spending on qualified research in the same year. For the research tax credit, there are two such measures: QREs, as reported by the IRS, and business spending on domestic basic and applied research and development, as reported by the National Science Foundation (NSF). As a result, the average effective rate can be computed for both QREs and business investment in domestic R&D.

As Table 3 shows, the average effective rate of the credit from 2000 to 2010 was 3.3% for business investment in domestic R&D and 5.3% for QREs. This implies that the credit lowered the average after-tax cost of that investment by 3.3% and of qualified research by 5.3% during that decade. By contrast, the statutory rate is 20% for the regular credit and 14% for the ASC. While claims for the regular credit accounted for nearly all QREs in 2000, QREs associated with claims for the ASC amounted to 67% of total QREs in 2010.

The gap between the rates largely reflects the difference between the scope of QREs and the scope of business R&D spending, as estimated by the NSF. Aggregate QREs amounted to 62.4% of aggregate business R&D

Table 3. Business and Federal Spending on Domestic Research and Development, and Claims for the Federal Research Tax Credit, 2000 to 2010 ($ billions)

	2000	*2001*	*2002*	*2003*	*2004*	*2005*	*2006*	*2007*	*2008*	*2009*	*2010*
Business Spending on Domestic R&D (BSDRD)[a]	$183	$185	$177.5	$183	$188	$204	$223	$243	$254	$243	$245
Qualified Research Spending (QRS)[b]	$110	$100	$116	$124.5	$116	$130	$145	$158	$151	$143	$160
Federal R&D Spending[c]	$76	$84.5	$94	$103.5	$112	$119	$122	$127	$127	$133	$147
Current-Year Research Tax	$7.2	$6.4	$5.7	$5.5	$5.6	$6.4	$7.3	$8.3	$8.3	$7.9	$8.5
Credit[d]											
Ratio of Credit to BSDRD (%)	3.9%	3.5%	3.2%	3.0%	3.0%	3.1%	3.3%	3.4%	3.3%	3.2%	3.5%
Ratio of Credit to QRS (%)	6.5%	6.4%	4.9%	4.4%	4.8%	4.9%	5.0%	5.2%	5.5%	5.5%	5.3%

Source: National Science Foundation, Division of Science Resources Statistics, Science and Engineering Indicators 2014, appendix table 4-3; National Science Foundation, Division of Science Resources Statistics, Federal Funds for Research and Development: Fiscal Years 2010-12, table 1; Internal Revenue Service, available at http://www.irs.gov/ uac/SOI-Tax-Stats-Corporation-Research-Credit.

[a] Total spending on domestic basic and applied research, as well as development, by companies only.

[b] Spending on research that qualifies for the regular, alternative incremental, and university basic research tax credits, as reported by corporations claiming the credit on their federal income tax returns.

[c] Federal obligations for defense and non-defense R&D spending by fiscal year.

[d] Total value of claims for the regular, incremental and basic research tax credits reported in federal corporate income tax returns. Because of limitations on the use of the general business credit, of which the research credit is a component, and audits of corporate claims for the credit by the Internal Revenue Service, the total amount of the research credit actually used in a particular year may differ from the total amount claimed.

spending from 2000 to 2010. The NSF estimate covers domestic R&D funded by firms. It is based on annual surveys of business R&D and takes into account the wages, salaries, and fringe benefits of research personnel; the cost of materials and supplies, overhead expenses; and depreciation related to research activities. Excluded from the estimate are expenditures on the buildings and equipment used in research, quality control, routine product testing, and prototype production.[24] By contrast, QREs represent eligible spending on research that qualifies for the credit, as reported to the IRS on Form 6765. Qualified expenses consist of the wages and salaries of research personnel, materials, supplies, leased computer time used in qualified research, and 65% to 75% of contract research funded by the taxpayers claiming the credit. The NSF figures are larger than the QREs for the same years because they cover more of the overall cost of business R&D investment.

The figures in Table 3 suggest that the credit delivered, at most, a modest stimulus to domestic business R&D investment from 2000 to 2010. Specifically, assuming the price elasticity of demand for qualified research lies between -0.4 and -0.8, and the credit lowers the net cost of business spending on qualified research by 5.3%, one can argue the credit may have boosted that spending somewhere between 2.1% (0.4 x 5.3%) and 4.2% (0.8 x 5.3%), relative to the investment that may have taken place in the absence of the credit.[25]

POLICY ISSUES RAISED BY THE CURRENT RESEARCH TAX CREDIT

Many policy analysts and lawmakers endorse the use of tax incentives to spur increased business R&D investment. Yet the current research tax credit sometimes seems to attract more words of criticism than praise. The main concern of critics is that the credit has not been as effective as it should have because of what they say are certain problems with its design. In their view, the credit can yield its intended benefits only if it is altered to remedy five problems in particular: (1) the credit is not a permanent provision of the IRC; (2) it still has weak and uneven incentive effects; (3) it is not refundable; (4) the definition of qualified research remains incomplete and ambiguous, and thus a major source of disputes between the IRS and taxpayers; and (5) the credit is not targeted at investments that are likely to generate relatively large economic benefits. Each potential difficulty is discussed in some depth below.

Lack of Permanence

The research tax credit expired on December 31, 2013. Several bills to extend it permanently have been introduced in the 113th Congress—a step that the Obama Administration has supported in its budget requests for FY2012, FY2013, and FY2014. The credit has never been a permanent provision of the IRC, despite repeated initiatives in Congress to extend it permanently in the past decade.[26] In fact, the credit has now been extended 15 times, most recently by the American Taxpayer Relief Act of 2012.

This lack of permanence is a cause for concern because there is reason to believe that it weakens the credit's incentive effect. Many R&D projects have planning horizons that extend beyond a year or two. If business managers cannot count on receiving the credit over the expected life of an R&D project, they may not take it into account when setting the size of annual R&D budgets. If this were to happen, the credit would be likely to exert little or no influence over R&D investment decisions. Instead of boosting R&D investment, a temporary research tax credit could end up restraining it by compounding the uncertainty that typically characterizes projected after-tax returns on planned R&D investments, and by keeping the cost of capital for such investments higher than it otherwise would be. A combination of added uncertainty and increased capital costs may deter managers from pursuing some of the R&D projects they would undertake if the credit were permanent.

Still, it seems unlikely that all firms investing in R&D would be affected in the same manner by a temporary research tax credit. Firms with relatively long R&D planning horizons and relatively high fixed costs for R&D investment might show more sensitivity to uncertainty in the availability of a credit than firms with shorter horizons and relatively low fixed investment costs. For example, it is conceivable (though difficult to verify) that a string of temporary credits may lead pharmaceutical firms to expand their research budgets more slowly than software firms, simply because pharmaceutical R&D projects, on average, have longer planning horizons and require greater investments in plant and equipment than software R&D projects do.

Uneven and Inadequate Incentive Effects

Critics maintain that another significant problem with the research tax credit is its incentive effect. A credit's incentive effect refers to the size of the benefit it offers eligible taxpayers. In the view of critics, the research credit's incentive effect varies among firms conducting qualified research in ways that are not supported by economic theory and can even defeat the credit's purpose. In addition, they contend that this incentive effect is too meager to offset the inclination of firms to invest less in research than its potential spillover benefits warrant. To what extent can it be said that the credit's incentive effect is too uneven and small?

Uneven Incentive Effect

The regular credit's incentive effect appears to vary widely among firms investing in qualified research, including those that gradually increase their research investment over an extended period. Evidence for such variation can be found in a number of sources, including a 1996 study by economist William Cox that estimated which of a large group of domestic corporations with sizable research budgets in 1994 should have be able to claim the regular credit.

The study was based on a sample of 900 publicly traded U.S.-based firms with the largest R&D budgets, culled from a database maintained by Compustat, Inc. Under the reasonable assumption that QREs for these firms were equal to 70% of their reported R&D spending for 1994, Cox determined that 62.5% of the firms could be considered established firms for the purpose of claiming the regular credit, as they had both business revenue and QREs in three of the years from 1984 to 1988; the remainder were treated as start-up firms. Cox found that 78% of the 900 firms in the sample (44.4% of the established firms and 33.5% of the start-up firms) could have claimed the credit in 1994, while 22% could have claimed no credit (18% of established firms and 4% of start-up firms).[27] He also found that 34% of all firms (32.3% of established firms and 1.7% of start-up firms) had QREs greater than their base amounts but less than twice those amounts, allowing them to claim credits with a marginal effective rate of 13%, and that 43.8% of all firms had QREs greater than double their base amounts, allowing them to claim credits with a marginal effective rate of 6.5%.[28] These rates measure the reduction in the after-tax cost of qualified research as a result of taking the regular credit. In addition, Cox found that some of the most research-intensive firms could claim either no credit or credits with a marginal effective rate half as large as the rate for the credits that could be claimed by firms with much lower propensity to invest in research.

The results seemed to confirm that the regular credit was most beneficial to firms whose research intensities had grown since their base periods and least beneficial to firms whose research intensities had changed little or not at all or had shrunk since their base periods. Most of the firms whose research intensities had declined found themselves in that position for two reasons: (1) their R&D spending was lower in 1994 than it was in their base period, or (2) their sales revenue had grown faster than their R&D expenditures over the same period.

Critics of the design of the regular credit argue that the pattern of R&D subsidization found in the Cox study is unfair and arbitrary, has no justification in standard economic theory, and undercuts the intended purpose of the credit, which is to encourage all firms to spend more on R&D than they otherwise would. Cox said as much when he concluded that the wide variation in the marginal effective rates of the credit among the firms in his analysis suggested "that society places a higher value on adding R&D at certain firms than at others and on adding R&D of certain types than others, when little or no basis for such different valuations exists."[29]

Two rules governing the use of the regular credit are responsible for most of the variation in its incentive effect. One is the requirement that the base amount for the regular credit cannot be less than 50% of QREs. The other rule is the requirement that established firms use gross receipts and QREs from 1984 to 1988 to calculate their fixed-base percentages.

In combination, the rules can produce remarkably dissimilar outcomes in the use of the regular credit among firms that spend substantial amounts on qualified research. Of particular concern to critics are firms whose research-intensity (as measured by spending on R&D as a share of revenue) has shrunk over time. The structure of the U.S. economy can and does change markedly in a period of 20 or so years. So it is likely that economic and competitive conditions in research-intensive industries today bear little resemblance to the conditions that prevailed in the 1980s. Most of the firms that have remained in business as independent entities and invested considerable amounts in R&D relative to revenues since then face much different climates for R&D investment. In some cases, the change in circumstances has led established firms to invest less in R&D as a share of revenues. Firms in this position may not be able to claim the regular credit, even if they spend relatively large sums on R&D.[30]

Inadequate Incentive Effect

In claiming that the regular credit's incentive effect is inadequate, critics actually have in mind two different measures of the effect. One deals with the credit rate deemed essential to coaxing companies to increase their R&D investments to socially optimal levels; the other concerns differences between the regular credit's statutory rate and its average marginal effective rate. Both measures are examined here.

Research Credit Rate and Socially Optimal Levels of Investment in Research

Critics maintain that the average effective rate of the regular credit is too low to support levels of business investment in research commensurate with its economic benefits. To substantiate this contention, they point to another study by Cox, one that focused on the efficacy of the research tax credit.[31] Cox built the analysis around the premise that tax incentives can overcome the private sector's inclination to invest too little in the creation of new technical knowledge and know-how. For tax incentives to have this effect, they must be designed so they subsidize R&D spending above and beyond what firms would undertake on their own, and they must be large enough to "raise private after-tax returns on R&D investments to the levels that would result from applying the same rate of taxation to the social rate of return from R&D."[32] A variety of studies from the past 50 years or so have concluded that the median private rate of return on R&D investment is roughly 50% of the median social rate of return.[33] Thus, assuming that the average social pre-tax rate of return is two times the average private pre-tax rate of return, the optimal R&D tax subsidy would double the private after-tax rate of return to R&D investment. For example, given a corporate tax rate of 35%, after-tax returns would equal 65% of pre-tax returns for corporations, assuming no tax subsidies or preferences. In this case, the optimal R&D tax subsidy would double the private after-tax returns to R&D investment by increasing them to 130% of pre-tax returns [2 x (1-0.35)].

Cox's analysis implied that the optimal average effective rate for an R&D tax subsidy, or a combination of such subsidies (e.g., a research tax credit combined with the treatment of research expenditures as a current business expense), was 30%. In discussing the policy implications of this finding, Cox noted that such a rate was an average and thus would not address the considerable variation among R&D investments in the difference between private and social returns. So using tax incentives to boost pre-tax returns on R&D investment by 30% across all industries would provide excessive subsidies for projects with below-average spillover benefits and insufficient subsidies for projects with above-average spillover benefits. According to Cox, lawmakers should be aware that "this imprecision is unavoidable, and its consequences are hard to assess."[34]

How do existing federal tax subsidies for R&D investment compare with Cox's assessment of the optimal R&D tax subsidy? To determine the incentive effect of then-current federal subsidies, he estimated the pre-tax and after-tax rates of return under 1995 federal tax law for a variety of hypothetical R&D projects. The projects differed in the share of R&D expenditures devoted to depreciable assets like structures and equipment, the share of R&D expenditures eligible for both expensing under IRC Section 174 and the regular research credit, and the economic lives of the intangible assets created by the investments. Cox compared the combined effect of expensing and the credit on after-tax returns to investment in capital-intensive, intermediate, and labor-intensive R&D projects producing intangible assets with economic lives of 3, 5, 10, and 20 years.[35]

Expensing has the effect of equalizing the pre-tax and after-tax rates of return on an investment, since it taxes the income earned by affected assets at a marginal effective rate of zero.[36] For the typical business R&D investment, it is likely that only part of the total cost may be expensed under IRC Section 174, as tangible depreciable assets like structures and equipment do not qualify for such treatment. Therefore, how expensing affects an R&D investment's after-tax rate of return depends on two factors: (1) the percentage of the total cost that may be expensed, and (2) the marginal effective tax rate on income earned by the assets (including labor) eligible for expensing.

The regular research credit raises the after-tax rate of return only on a portion of current-year QREs: those above a base amount. So its effect on the after-tax returns to an R&D investment depends on both the percentage of the investment's total cost that qualifies for the credit and the effective tax rate on income earned by assets eligible for the credit.

Allowing for these limitations on the benefits of expensing and the regular credit, Cox estimated that expensing and the credit together produced median after-tax rates of return ranging from 101.0% of pre-tax returns for a hypothetical capital-intensive project yielding intangible assets with an economic life of 20 years to 124.7% for a hypothetical labor-intensive project yielding intangible assets with an economic life of three years.[37] As these percentages are less than 130%, he inferred that the research tax subsidies in existence in 1995 did not increase private after-tax returns to R&D investments to the "levels warranted by the spillover benefits that are thought to be typical" for these investments.[38]

Gap Between the Credit's Average Effective Rate and Its Statutory Rate

Some critics of the regular credit see its incentive effect in a somewhat different, though related, light. For them, a more important concern than the credit's potential to boost business investment in research to socially optimal levels is any differences between the regular credit's average effective rate and its statutory rate of 20%. As noted earlier, whatever differences exist from firm to firm can be ascribed largely to three of the rules governing the use of the credit.

One of the rules is the basis adjustment under IRC Section 280C(c)(1), which requires taxpayers investing in qualified research to reduce any deduction for research expenditures they take under IRC Section 174 by the amount of any research credit they claim. This adjustment effectively adds the credit to a firm's taxable income and taxes it at the firm's income tax rate. Consequently, for business taxpayers taking the credit that are subject to the maximum corporate or individual tax rate of 35%, the basis adjustment decreases the marginal effective rate of the credit from 20% to 13%. Firms have the option of computing the regular research credit at a statutory rate of 13% in exchange for leaving intact any Section 174 deduction.

A second rule is the requirement that the base amount for the regular credit cannot be less than 50% of a firm's current-year QREs. Perhaps unintentionally, the rule curtails the credit's potential benefit to established firms whose ratio of current-year QREs to gross income is more than double their fixed-base percentages, or more than double the 16% cap on the fixed-base percentage—which is to say established firms that historically have invested heavily in qualified research. Start-up firms, whose current-year ratio of QREs to gross income exceeds 6% during their first five tax years, or whose current-year ratio is more than double their fixed-base percentages in the next six tax years, also are affected by the rule. For both sets of firms, it further reduces the marginal effective rate of the credit to 6.5%.

The third rule is the exclusion of expenditures for equipment and structures and overhead costs from expenses eligible for the regular credit—even though many business research investments involve the acquisition of elaborate buildings and sophisticated equipment, and all research projects have overhead costs. In this case, the rule's effect on the marginal effective rate of the credit depends on the share of the cost of an R&D investment to which the credit does not apply. As this share rises, the credit's marginal effective rate drops, all other things being equal. For example, if expenditures for physical capital account for half of the cost of an R&D investment, then the marginal effective rate of the credit for the entire investment is half of what it would be if the entire cost were eligible for the credit. For firms subject to the 50% rule that invest in research projects where physical capital represents 50% of the total cost, the marginal effective rate could fall to 3.25%.

In addition, in a 2009 report on problems with the credit's design, the GAO brought into the analysis of the credit's marginal effective rate the impact of delays in the use of the credit. In essence, they reduce the present value of the credit, and such a reduction lowers the rate. The longer the delay and the larger a taxpayer's discount rate, the larger the rate decline. GAO estimated the marginal effective rate for all the corporations in the IRS database that claimed the credit in 2003 to 2005 and used them to compute a weighted average rate for all taxpayers. It found that the rate ranged from 6.4% to 7.3%, depending on the assumptions about the length of any delay in using the credit and the discount rate.[39]

As these considerations suggest, one of the keys to bolstering the regular credit's incentive effect is to increase its marginal effective rate. There are two ways to do so: (1) keep the current statutory rate and relax or repeal one or more of the three rules, and (2) retain the rules but raise the credit's statutory rate to offset their effects.

Cox analyzed the effects of both options on after-tax rates of return for the same hypothetical R&D investments discussed above. In the case of labor-intensive R&D projects, he estimated that 1995 research tax preferences produced median after-tax returns that were 124.7% of pre-tax returns for projects yielding intangible assets with an economic life of three years, and 115.5% for projects yielding intangible assets with an economic life of 20 years. Repealing the basis adjustment for the credit caused median after-tax returns to increase to 146.0% of pre-tax returns for assets with a three-year economic life, and 130.1% for assets with a 20-year economic life.[40] Increasing the statutory rate of the credit to 25% but retaining existing rules (including the basis adjustment) led to similar results: median after-tax returns for assets with a three-year economic life were an estimated 133.9% of pre-tax returns, and an estimated 121.9% for assets with a 20- year economic life.[41] As one might expect, increasing the rate to 25% and removing the basis adjustment led to the biggest boost in the ratio of after-tax returns to pre-tax returns: 165.8% for assets with a three-year economic life, and 143.4% for assets with a 20-year economic life.

If it is true that the optimal R&D tax subsidy should raise after-tax returns to 130% of pre-tax returns, Cox's analysis suggested that keeping the regular credit's statutory rate at the current level of 20% but eliminating or relaxing the three rules governing the credit's use might be the better policy option for strengthening the credit's incentive effect.

Non-refundable Status

The research tax credit is non-refundable. This means that only firms with sufficient income tax liabilities may benefit from the full amount of the credit claimed in a tax year. In addition, the credit is a component of the general business credit (GBC) under IRC Section 38, and therefore subject to its limitations. For firms undertaking qualified research, a key limitation is that the GBC cannot exceed a taxpayer's net income tax liability, less the greater of its tentative minimum tax under the alternative minimum tax or 25% of its regular income tax liability above $25,000. Unused GBCs may be carried forward 20 years or back one year. Although there are some advantages to having an inventory of tax credits to apply against future or past tax liabilities, the advantages do not necessarily outweigh the disadvantages for firms investing in R&D. One disadvantage is that a business taxpayer is better off using the full amount of a credit today, rather than 5 or 10 years from now, owing to the time value of money.

Critics contend that the credit's lack of refundability can pose a special problem for small, fledgling, research-intensive firms. In recent decades, numerous commercially successful technological innovations have originated with such firms. Many of them spend substantial sums on R&D during their first few years, while recording a stream of net operating losses. In the view of critics, a non-refundable research credit is likely to reduce the typical small, research-intensive start-up firm's prospects of survival and growth, as the firm cannot count on having access to the credit when the need for it is greatest. To remedy this problem, some advocate making the credit wholly or partially refundable for firms under a certain asset or employment size or age.[42] Other options include allowing small start-up firms that cannot use the current-year credit to sell it to other firms or use it to offset their employment taxes.[43]

Incomplete and Ambiguous Definition of Qualified Research and Difficulties in Claiming the Credit

Critics claim that another problem raised by the current research tax credit is the activities that qualify for it. At its core, the issue concerns the definition of qualified research and how the IRS and taxpayers interpret it in the real world of business investment in basic and applied research.

Critics argue that the statutory definition in IRC Section 41(d) and IRS regulations implementing it are vague and incomplete. In their view, this lack of clarity and finality, coupled with the complexity of the credit, the difficulty of documenting claims for the credit in a manner that passes muster with the IRS, and a lack of useful guidance from the IRS and courts, paves the way for protracted and costly disputes between taxpayers and the IRS over the validity of claims for the credit.[44] These disputes can dampen the stimulative effect of the credit in two ways. IRS audits of claims for the credit can lead to firms receiving smaller credits than they thought they were entitled to. In addition, the prospect of having a claim audited and engaging in a lengthy dispute with the IRS over its legitimacy can deter some firms that invest in qualified research from claiming the credit.

Original Definition

Under the original credit, which was in effect from 1981 through 1985, research expenditures generally qualified for the credit if they were eligible for expensing under IRC Section 174. There were three exceptions to this general rule, however: no credit could be claimed for research conducted outside the United States, research in the social sciences or humanities, and any portion of research funded by another entity. Section 174 allows business taxpayers to deduct all "research or experimental expenditures" incurred in connection with their trade or business, without defining those outlays.

The IRS filled the gap by issuing regulation 1.174-2(a), which defined research or experimental expenditures as "research and development costs in the laboratory sense," especially "all such costs incident to the development or improvement of a product." Expenditures can be considered R&D costs in the "experimental or laboratory sense" if they relate to activities intended to discover information that would eliminate uncertainty concerning the development or improvement of a product. Uncertainty exists in the R&D process when the information available to researchers does not clearly show how they should proceed in developing a new product or improving an existing one. According to the regulation, the proper standard in determining whether research expenditures qualify for expensing under Section 174 is the "nature of the activity to which the expenditures relate, not the nature of the product or improvement being developed."

Changes Under the Tax Reform Act of 1986

Responding to a concern that business taxpayers were claiming the credit for activities that had more to do with product development than technological innovation, Congress tightened the definition by adding three tests through the Tax Reform Act of 1986 (TRA86).[45] Under the act, qualified research still had to involve the activities eligible for expensing under Section 174. But such activities also had to satisfy the following criteria:

- they were directed at discovering information that is "technological in nature" and useful in the development of a new or improved business component for the taxpayer;
- they constituted "elements of a process of experimentation"; and
- they were intended to improve the function, performance, quality, or reliability of a business component.[46]

TRA86 defined a business component as "a product, process, computer software, technique, formula, or invention" held for sale or lease or used by a taxpayer in its trade or business. It also specified that research aimed at developing new or improved internal-use software could qualify for the credit only if it met the general requirements for the credit, was intended to develop software that was innovative and not commercially available, and involved "significant economic risk."

Subsequent IRS Guidance

In light of the significant changes made by the act, there was renewed pressure on the IRS to issue final regulations clarifying the meaning and limits of the three new tests for qualified research. But for reasons that still are not entirely clear, the IRS did not issue proposed regulations (REG-105170-97) on the tests until December 1998.

Among other things, the regulations set forth guidelines for determining whether or not a business taxpayer has discovered information that is "technological in nature" and "useful in developing a new or improved business component of the taxpayer" through a "process of experimentation that relates to a new or improved function, performance, reliability, or quality." The IRS proposed that research would meet the so-called discovery test if it were intended to obtain "knowledge that exceeds, expands, or refines the common knowledge of skilled professionals in the particular field of technology or science." At the same time, according to the proposed regulations, such a standard did not necessarily mean the credit would be denied to taxpayers who made technological advances in an "evolutionary" manner, or taxpayers who failed to achieve the desired result, or taxpayers who were not the first to achieve a certain technological advance. In addition, the IRS proposed that research would meet the experimentation test if it were to draw upon the "principles of physical or biological sciences, engineering, or computer science (as appropriate)" to evaluate "more than one alternative designed to achieve a result where the means of achieving the result are uncertain at the outset." Such an evaluation should involve developing, testing, and refining or discarding hypotheses related to the design of new or improved business components.

The release of the proposed regulations drew more criticism than praise from the business community. Much of the dissent focused on the proposed guidelines for the discovery test. A widely shared complaint was that the "common knowledge" test violated the intent of Congress and would prove burdensome and unworkable for tax practitioners because it was too subjective. Most tax practitioners and businesses that commented on the proposal urged the IRS to jettison the test.[47]

After reviewing the many comments it received and examining recent case law and the legislative history of the research tax credit, the IRS issued what was intended to be a final set of regulations (T.D. 8930) on the definition of qualified research in late December 2000. While differing somewhat from the proposed regulations, the final regulations retained the common knowledge test for determining if the information gained through research was technological in nature and useful in the development of a new or improved business component. But they clarified the application of the test by noting that the "common knowledge of skilled professionals in a particular field of science or engineering" referred to information that would be known by those professionals if they were to investigate the state of knowledge in a field of science or engineering before undertaking a research project. The final regulations also carved out a safe harbor for patents by affirming that a taxpayer would be presumed to have passed the common knowledge test if the taxpayer could prove it had been awarded a patent for a new or improved business component. They also established new standards for determining when the development of computer software for internal use qualified for the credit. Specifically, research on internal-use software was eligible for the regular credit only if it satisfied the general requirements for the credit, entailed "significant economic risk," and resulted in the development of innovative software that was not commercially available.

To the surprise of some, the final regulations aroused as much opposition within the business community as the proposed regulations. A principal concern was the IRS's insistence on retaining the discovery test. Many tax practitioners also complained that a number of the provisions in the final regulations were not included in the proposed regulations, precluding public comment on them.[48]

This second round of criticism spurred the IRS to take an unusual procedural step. About one month after the release of the regulations, the Treasury Department published a notice (Notice 2001-19) retracting them. The notice also requested further comment "on all aspects" of the suspended regulations, promised that the IRS would carefully review all questions and concerns, and committed the IRS to issue any changes to the final regulations in proposed form for additional comment.[49]

In December 2001, the IRS delivered on this promise by releasing more proposed regulations (REG-112991-01). They departed in some important ways from previous guidance. Among other things, the regulations did not include the requirement set forth in T.D. 8930 that qualified research seek to discover "knowledge that exceeds, expands, or refines the common knowledge of skilled professionals in a particular field of science or engineering." The regulations also modified the definition of the experimentation test so that it became a "process designed to evaluate one or more alternatives to achieve a result where the capability or the method of achieving that result, or the appropriate design of that result is uncertain as of the beginning of the taxpayer's research activities." The determination of whether a taxpayer engaged in such a process would be made on the basis of relevant facts and circumstances. In addition, the proposed regulations stipulated that internal-use software could not to be sold, leased, or licensed to third parties and was eligible for the credit only if it is intended to be novel in its design or applications. Tax practitioners and businesses generally approved the proposed changes.[50]

About two years later, the IRS published another set of final regulations (T.D. 9104) that was intended to clarify the definition of qualified research and certain other matters related to the credit.[51] Some analysts saw them as an attempt by the IRS to adhere to congressional intent in altering the definition of qualified research, as specified in TRA86.

The regulations noted that information is technological in nature if the process of experimentation used to discover it relies on the principles of the physical or biological sciences, engineering, or computer science. Though they discarded the discovery test included in T.D. 8930, the regulations made it clear that taxpayers would be deemed to have discovered information that is technological in nature by applying "existing technologies.... and principles of the physical or biological sciences, engineering, or computer science" in the process of experimentation. Such a discovery would not depend on whether a taxpayer succeeded in developing a new or improved business component. At the same time, having a patent for a business component would be considered "conclusive evidence that a taxpayer has discovered information that is technological in nature that is intended to eliminate uncertainty concerning the development or improvement of (such a) component."

In addition, T.D. 9104 shed additional light on what constituted a "process of experimentation." Basically, the regulations specified that such a process had three critical aspects. First, the actual outcome must be uncertain at the outset. Second, the process must allow researchers to identify more than one approach to achieving a desired outcome. And third, researchers must use certain scientific methods to evaluate the efficacy of these alternatives (e.g., modeling, simulation, and a systematic trial-and-error investigation). The regulations stressed that a process of experimentation is evaluative in nature and therefore "often involves refining throughout much of the process the taxpayer's understanding of the uncertainty the taxpayer is trying to address." A taxpayer's relevant facts and circumstances should be considered in determining whether it has engaged in such a process.

Remaining Concerns

Although the regulations seemed to settle several contentious issues regarding the definition of qualified research, they did not address several other issues regarding claims for the credit that are viewed as consequential by many firms.

One issue is the circumstances under which spending on the development of internal-use software can be deemed eligible for the credit. In proposed regulations issued in 2001, the IRS stated that any costs incurred to develop such software were eligible for the credit only if the software was intended to be unique or novel and to differ in a "significant and inventive" way from previous software. But in the absence of further guidance on the meaning of "significant and inventive," disputes between IRS examiners and taxpayers over the validity of claims for the credit tied to internal-use software are more likely than not. One analyst has noted that since the release of the final regulations, the IRS has interpreted the definition of significant and inventive in a way that imposes the same requirements on the development of internal-use software that the discarded discovery test did.[52]

Another issue is the eligibility of research aimed at achieving significant cost reductions. Cost reduction is not identified in the statute as a purpose of qualified research, but the research required to lower costs can be as challenging as research done to improve a business component's reliability or performance. Some have pointed out that research that allows a product or process to deliver the same performance at a reduced cost represents an improvement in performance.[53]

In addition, tax practitioners have complained that in the wake of the strong opposition within the business community to T.D. 8930, the IRS has stopped using revenue rulings and regulations in favor of internal administrative announcements to communicate its interpretations of the IRC Section 41. These announcements, according to these critics, are sent to IRS employees in an effort to build a uniform approach to auditing claims for the credit that avoid public comment. Critics charge that these internal communications force some business taxpayers to "resort to costly litigation to determine the application of Section 41 to their facts and circumstances," undermining the credit's incentive effect.[54]

Finally, substantiation of claims for the credit remains a significant source of disputes with the IRS, curtailing its incentive effect. Concerned about the legitimacy of late or amended claims for the credit, the IRS decided to designate "research credit claims" as a tier 1 compliance issue in April 2007. Supporting claims for the credit typically requires subjective judgments that are subject to differing interpretations. These differences often surface in IRS examinations of claims for the credit. Examiners allegedly "routinely" deny claims because they do not strictly follow IRS's documentation standards, even though the standards never have been clarified in IRS regulations.[55] Taxpayers are upset because the IRS expects them to provide supportive evidence without useful guidance on the required documentation standards. In its defense, the IRS maintains that it cannot audit claims for the credit unless it has detailed records linking specific expenses to specific qualified research activities. According to a 2009 GAO report, in many cases, "taxpayers settle for 50 cents on the dollar when the IRS challenges a claim."[56]

Insufficient Focus on Innovative Research Projects

Another problem with the current research credit, according to critics, is that it is not targeted at research projects with a significant potential for producing substantial economic benefits. In their view, the design of the credit makes it likely that it generates less bang for the buck in the long run than federal grants for basic and applied research in fields with broad commercial applications.

Assessing the validity of this claim is a challenge, largely because it is unclear at the outset of a research project how innovative it will prove to be. There are no known studies that undertake such an assessment. Nonetheless, the contention seems plausible in a way that might be difficult to disprove.

In general, businesses seek the highest possible return on their investments. So in selecting research projects to fund, they can be expected to assign a higher priority to projects that are likely to earn substantial profits in the short run than to projects directed at expanding the frontiers of knowledge in a scientific field that have relatively low prospects of yielding profits in the short run. This assumes, of course, that a firm has the option of investing in either kind of project or both at the same time.

Such an expectation not only seems reasonable by the logic of business investment and growth, but it is consistent with several significant trends in U.S. research spending that stretch back to the 1950s. As *Figure 1* illustrates, the federal government has long served as the major source of funding for basic research performed in the United States; from 1955 to 2008, its share of total spending (in current dollars) for this purpose was about three times the share for businesses, though the gap has narrowed somewhat since the early 1980s. At the same time, U.S.- and foreign-based companies steadily expanded their share of total funding for applied research and development performed in the United States in that period; by 2008, the business share was 88% greater than the federal share for applied research and more than five times greater for development.

These trends arguably confirm that the average firm prefers to invest much more in applied research and development than in basic research. This is hardly surprising, as the returns on investment in basic research tend to be more difficult to appropriate and more uncertain at the outset, than the returns to investment in applied research and development. The trends also lend support to the view that the credit mainly subsidizes research projects with relatively small spillover benefits, as it applies to qualified projects involving basic and applied research, as well as development.

To address this concern, critics recommend that Congress modify the credit so that it expressly targets investment in research aimed at developing "breakthrough products that create new product categories or innovative enhancements to existing products."[57] Among the options are (1) replacing the current credit with one that has a larger credit rate (say 30%) for such expenditures and (2) transferring the authority to administer the credit to the

NSF. A presumed advantage of using the NSF rather than the IRS to identify investments that would qualify for the revised credit is that the NSF has a greater supply of the expertise required to evaluate research projects for the purpose of determining which could be regarded as innovative.[58]

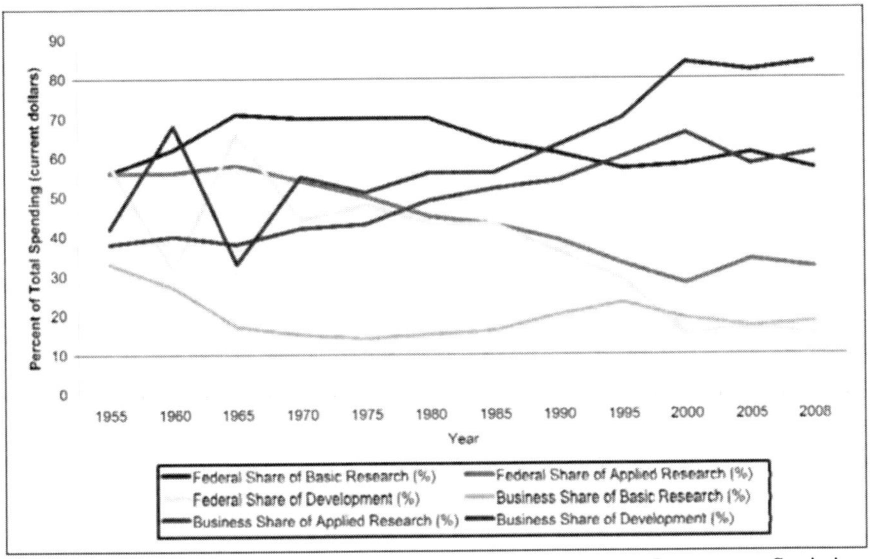

Source: National Science Foundation, Division of Science Resources Statistics, National Patterns of R&D Resources: 2008 Data Update, NSF 10-314, tables 6-8.

Figure 1. Share of U.S. Spending (current dollars) on Research and Development Held by the Federal Government and Businesses, 1955 to 2008.

LEGISLATION IN THE 113TH CONGRESS TO MODIFY AND EXTEND THE RESEARCH TAX CREDIT

It is fair to say that the research tax credit has enjoyed strong bipartisan support since its inception, and the current Congress is no exception.

Yet in spite of that backing, the credit remains temporary and less effective than it might be. For over two decades, a major obstacle to extending the credit beyond a year or two and to improving its efficacy has been the revenue cost of doing so. If anything, this issue may be even more of an obstacle in the current Congress, which has been struggling to find politically acceptable ways to reduce or eliminate current and projected future budget deficits and the federal debt.

Proposals in the current Congress related to the research tax credit range from a simple one-year extension, through 2014, to permanent extensions with an increase in the rate for the ASC to 20% and the option for eligible small companies to use the credit to reduce their payroll taxes.

113th Congress

House

H.R. 119 would permanently extend the research tax credits.

H.R. 120 would extend the credits through 2014, raise the rate of the regular credit to 30% and the rates for the maximum and minimum ASC to 20% and 12% in 2013 and 2014 only, and permit qualified small firms to transfer credits they can claim but not use to other businesses for cash and to exclude those payments from their gross incomes.

H.R. 714 would modify the Section 41 credit to allow qualified small firms to apply a portion of any research tax credit they claim but cannot use because of insufficient income tax liability against their federal payroll taxes; the portion that can be used for this purpose would be limited to $250,000 in a tax year.

H.R. 905 would permanently extend the research tax credits and raise the rate for the ASC to 17%.

H.R. 3355 would extend the Section 41 credit for one year, through the end of 2014, permanently extend the ASC, and raise its rate to 20%.

H.R. 3428 would extend the Section 41 credit through 2018 and raise the rate for the ASC to 20%. It would also establish a new separate credit for expenditures on development activities under Section 41A, which also would be available through 2018. The credit would equal 30% of QREs above 50% of a company's average annual QREs in the three previous tax years; for companies with fewer than three past years with QREs, the credit would equal 12% of current-year QREs.

H.R. 3537 would permanently extend the Section 41 credit and raise the rate of the credit to 25% for payments for contract research performed in the United States by manufacturers. The bill would also allow companies that earn income from the domestic production of property that was "substantially" developed in the United States to claim a Section 199 deduction of 15% instead of the 9% permitted under current law.

H.R. 3640 would permanently extend the Section 41 credit, raise the rates for the regular, university basic research, and energy research credits to 30%

(without changing the rate for the ASC), and allow eligible small businesses to apply the Section 41 credit to their payroll taxes, without limit.

H.R. 3757 would extend the Section 41 credit one year, through 2014, permanently extend the ASC and increase its rate to 20%, and allow the full amount of payments for contract research performed by biotechnology research consortia to be eligible for the credit, with the credit capped at 20% of such payments.

H.R. 3945 and *H.R. 4469* would extend the current credit one year, through 2014.

H.R. 4473 would permanently extend the current credit and raise the rate for the ASC to 20%.

H.R. 4696 would modify the credit so that small startup businesses could use a portion of any credit they can claim but cannot use because of insufficient income to offset their payroll taxes.

The House has passed two bills (*H.R. 4438,* the American Research and Competitiveness Act of 2014, on May 9 and H.R. 4, the Jobs for America Act, on September 18) to extend permanently and simplify the research tax credit. Under each measure, the credit would be equal to the sum of the following elements:

- 20% of a taxpayer's QREs in the current tax year above a base amount equal to 50% of its average annual QREs in the previous tax years,
- 20% of a taxpayer's payments in the current tax year for basic research done by a qualified organization under a written contract above a base amount equal to 50% of its average annual basic research payments in the three previous tax years, and
- 20% of the amount paid or incurred by a taxpayer in the current tax year for energy research conducted by an energy research consortium.

In the case of taxpayers that have no QREs or basic research payments in one or more of their three previous tax years, the credit is equal to 10% of their current-year QREs and basic research payments.

The Joint Committee on Taxation estimates that these changes in the credit would lead to a revenue loss of $155.5 billion from FY2014 to FY2024.[59] Under the bill, this cost would not be offset. While supporting a permanent extension of the credit, the Obama Administration has said it would veto the bill on the grounds that it would add too much to the projected federal deficit.[60]

Senate

S. 193 and *S. 310* would make the same change in the research tax credit as H.R. 714 would.

S. 1859 would extend the Section 41 credit through 2014, along with a number of other expired tax provisions affecting individuals and companies.

S. 1920 would permanently extend the current credit and allow it to be used in full against the alternative minimum tax. The bill would also establish a permanent 5-year carryback period and 24-year carryforward period for eligible small businesses that cannot use the full amount of any Section 41 credit they claim for the current tax year. In addition, the bill would allow companies to claim the ASC on amended tax returns and eligible small businesses to apply the credit against their payroll taxes, with a cap of $250,000 on the amount of the credit that could be used for this purpose. To be eligible for this option, the gross receipts in the current tax year for a C corporation, S corporation, partnership, or self-employed individual must be less than $5 million, and in the period preceeding the previous five tax years, the firm had no gross receipts.

S. 2715 would make several significant changes to the credit. First it would permanently extend the ASC and raise its rate to 25% for companies with QREs in the three previous tax years; the rate would be set at 10% for companies without QREs in any of their three previous tax years. Second, it would repeal the current regular, energy, and basic research credits. Third, S. 2715 would allow a company that acquires another company that performed qualified research to include a portion of the acquired company's QREs during the three tax years preceding the year of the acquisition in the acquiring company's QREs for the purpose of determing the base amount for the credit in the tax year that includes the acquisition. In addition, under the bill, the treatment of contract research expenses would be modified so that the full amount of any payments for contract basic research would be eligible for the credit. The research would have to be performed under a written contract by a university, a tax-exempt organization dedicated to scientific research, or an organization dedicated to making grants to universities for basic research. And S. 2715 would create an "enhanced credit for highly innovative research" under Section 41. The credit would be equal to 35% of QREs related to the development of a "qualified new product category" or a product that "represents a significant improvement over previously existing product technology."

On April 3, the Senate Finance Committee passed a bill (*S. 2260*, the Expiring Provisions Improvement, Reform, and Efficiency Act) that would extend the vast majority of the individual and business tax preferences that expired at the end of 2013. One of the preferences is the Section 41 research tax credit.

S. 2260 would make three changes in the expired credit. First, it would retroactively extend the credit through 2015. Second, it would allow the full amount of the credit to be applied to the corporate alternative minimum tax. Third, it would permanently alter the design of the credit to permit qualified companies that can claim the credit but cannot use some or all of it because of insufficient income tax liability in the current tax year to apply up to $250,000 of the credit against their current payroll tax liabilities. Companies that have been in business less than five years and that have less than $5 million in gross receipts would be eligible to take advantage of this option. Eligible companies with less income tax liability than the credit they can use would be allowed to split the credit between their income and payroll tax liabilities; the income tax liability would have to be offset first. The option would be available for tax years starting in 2014.

President Obama's Budget Proposal for FY2015

In its budget request for FY2015, the Obama Administration asks Congress to permanently extend the existing research credits and increase the rate for the ASC from 14% to 17%. The former change would go into effect for tax years beginning in 2014 and the latter for tax years beginning in 2015. The projected 10-year revenue cost for these changes is $108.1 billion from FY2015 to FY2024.[61]

Camp Tax Reform Proposal

The discussion draft for a tax reform bill issued by Representative Dave Camp in late February 2014 included a provision that would modify and permanently extend the research tax credit.

Under the proposal, the credit would be equal to 15% of current-year QREs above 50% of a company's average annual QREs in the previous three tax years plus 15% of basic research payments for the current tax year above 50% of the company's average annual basic research payments in the same period. Companies with fewer than three previous tax years would be able to claim a reduced credit, assuming they otherwise qualify.

The proposal would make several other significant changes to the current credit. First, it would repeal the 20-percent regular credit, the university basic research credit, and the energy research credit. Second, it would no longer allow amounts spent on computer software and materials and supplies to qualify for the credit. And finally, the proposal would eliminate all the differences under current law in the percentage of contract research payments that qualifies for the credit. As a result, 65% of all such payments would be eligible for the credit, regardless of which organization conducts the qualified research.

According to an estimate by the JCT, the proposal would have a revenue cost of $34.1 billion from FY2014 to FY2023.[62]

End Notes

[1] Linda R. Cohen and Roger G. Noll, "Privatizing Public Research," Scientific American, September 1994, p. 72.

[2] See, for example, Edwin Mansfield, "Microeconomics of Technological Innovation," in The Positive Sum Strategy, Ralph Landau and Nathan Rosenberg, eds. (Washington: National Academy Press, 1986), pp. 307-325; and John C. Williams and Charles I. Jones, "Measuring the Social Return to R&D," Quarterly Journal of Economics, vol. 113, no. 4, November 1998, pp. 1119-1135.

[3] For a brief discussion of these channels, see Bronwyn H. Hall, "The Private and Social Returns to Research and Development," in Technology, R&D, and the Economy, Bruce L. R. Smith and Claude E. Barfield, eds. (Washington: Brookings Institution and American Enterprise Institute, 1996), pp. 140-141.

[4] For more information on the Section 174 expensing allowance, see U.S. Congress, Senate Committee on the Budget, Tax Expenditures, committee print, 107th Cong., 2nd sess. (Washington: GPO, 2002), pp. 55-58.

[5] Office of Management and Budget, Analytical Perspectives, Fiscal Year 2014 (Washington: GPO, 2013), p. 374.

[6] Firms investing in qualified research that could not claim the regular credit had the option of taking what was known as an alternative incremental R&E tax credit (or AIRC), under IRC Section 41(c)(4), for tax years from 1996 to 2008. The Emergency Economic Stabilization Act of 2008 (P.L. 110-343) repealed the AIRC for the 2009 tax year, and Congress has not reinstated it.

[7] See the discussion of concerns raised by the current definition of qualified research in the "Incomplete and Ambiguous Definition of Qualified Research" section of this report.

[8] U.S. Office of Technology Assessment, The Effectiveness of Research and Experimentation Tax Credits (Washington: 1995), p. 29.

[9] For a variety of reasons, which will be discussed in a later section of the report, the actual or effective rate of the credit is much lower than 20%.

[10] Available studies indicate that the price elasticity of demand for R&D ranges from 0.2 to 2.0, which means that a 1% reduction in the cost of R&D would raise R&D spending between 0.2% and 2%.

[11] In other words, the expenses against which the regular research credit may be claimed can equal no more than 50% of total QREs in a given tax year.

[12] The definition of a start-up firm has changed a few times since the research credit was enacted. Presently, it denotes a firm that recorded gross receipts and QREs in a tax year for the first time after 1993.

[13] Calculating a firm's base amount for the basic research credit is more complicated than calculating its base amount for the regular credit. For the basic research credit, a firm's base period is the three tax years preceding the first year in which it had gross receipts after 1983. The base amount is equal to the sum of a firm's minimum basic research amount and its maintenance-of-effort amount in the base period. The former is the greater of 1% of the firm's average annual in-house and contract research expenses during the base period, or 1% of its total contract research expenses during the base period. For a firm claiming the basic research credit, its minimum basic research amount cannot be less than 50% of the firm's basic research payments in the current tax year. The latter is the difference between a firm's donations to qualified organizations in the current tax year for purposes other than basic research and its average annual donations to the same organizations for the same purposes during the base period, multiplied by a cost-of-living adjustment for the current tax year.

[14] U.S. Congress, Joint Committee on Taxation, General Explanation of the Economic Recovery Tax Act of 1981, joint committee print, 97th Cong., 1st sess. (Washington: GPO, 1981), p. 120.

[15] U.S. Congress, General Explanation of the Tax Reform Act of 1986, joint committee print, 100th Cong., 1st sess. (Washington: GPO, 1987), p.130.

[16] For a business taxpayer in the 30% tax bracket, the rule reduced the maximum effective rate of the regular research credit from 20% to 17.5%: .20 x [1 - (.5 x .30)].

[17] See U.S. Congress, Joint Economic Committee, The R&D Tax Credit: An Evaluation of Evidence on Its Effectiveness, joint committee print, 99th Cong., 1st sess. (Washington: GPO, 1985), pp. 17-22.

[18] Martin A. Sullivan, "Research Credit Hits New Heights, No End in Sight," Tax Notes, vol. 94, no. 7, February 18, 2002, p. 801.

[19] The principal barriers to measuring the social returns to R&D are developing adequate price indices for the cost elements of R&D for specific industries, specifying the time period in which to assess the productivity gains from R&D, and determining the depreciation rate for a society's stock of R&D assets. For a detailed discussion of these issues, see Bronwyn H. Hall, "The Private and Social Returns to Research and Development," in Technology, R&D, and the Economy, Bruce L. Smith and Claude E. Barfield, eds. (Washington: Brookings Institution, 1996), pp. 141-145.

[20] This argument assumes that government research grants to the private sector do not lead firms receiving the grants to reduce their own R&D spending by similar amounts.

[21] See Bronwyn H. Hall, Effectiveness of Research and Experimentation Tax Credits: Critical Literature Review and Research Design, report for Office of Technology Assessment, June 15, 1995, pp. 11-13, available at http://emlab.berkeley.edu/~bhhall/papers/ BHH95%20OTArtax.pdf.

[22] Ibid., p. 18.

[23] U.S. Congress, Joint Committee on Taxation, Description of Revenue Provisions in the President's Fiscal Year 2004 Budget Proposal, JCS-7-03, 108th Cong., 1st sess. (Washington: March 2003), p. 250.

[24] National Science Foundation, Division of Science Resource Statistics, The Methodology Underlying the Measurement of R&D Expenditures: 2000 (data update) (Arlington, VA: December 10, 2001), p. 2.

[25] This estimate assumes that all the credits claimed in each year of that decade were used immediately and were not subject to reduction because of IRS audits. Delays in the use of any credit shrink its present value, and thus its marginal effective rate. For instance, if a taxpayer claims a research credit of $1 million, has a discount rate of 5%, and cannot use the credit for three years, then the present value of the credit drops to about $864,000. Because the top marginal effective rate for the credit is 13%, owing to the rule that any deduction of research expenditures under Section 174 must be reduced by the amount of the credit, the delay in using the credit lowers the marginal effective rate to 11.2% (13% x 0.864). Delays can occur for two reasons: IRS audits or insufficient or no tax liability against which to apply credit in the current tax year.

A 2009 report on the design of the research tax credit by the Government Accountability Office (GAO) casts doubt on the plausibility of the assumptions underlying the estimated stimulative effect of the credit from 1998 to 2008. The report found that 44% of the "total net credits earned" in 2005 could not be used immediately and thus could be carried back one tax year or forward up to 20 tax years. It also noted that IRS data on examinations of claims for the credit by corporations with annual business receipts of $1 billion or more from 2000 to 2003 indicated that the examiners recommended changes that would have lowered the aggregate amount of credits claimed by 16.5% in 2000, 19.3% in 2001, 27.1% in 2002, and 24.5% in 2003. See U.S. Congress, Government Accountability Office, Tax Policy: The Research Tax Credit's Design and Administration Can Be Improved, GAO-10-136 (Washington: November 2009), pp. 14-15.

[26] The R&E tax credit has been in effect for each year between July 1, 1981, and the present except for period from July 1, 1995, to June 30, 1996, when it expired. Since July 1, 1996, the credit has not been renewed to include this period.

[27] CRS Report 96-505, Research and Experimentation Tax Credits: Who Got How Much? Evaluating Possible Changes, by William A. Cox, pp. 5-10. (The report is out of print. Copies may be obtained from Gary Guenther (202) 707-7742, upon request.) (Hereinafter cited as Cox, Research and Experimentation Tax Credits.)

[28] Their effective credit rate was lower because each firm was subject to the 50-percent rule, which reduced the marginal effective rate of the credit on R&D spending above the base amount by 50%.

[29] Cox, Research and Experimentation Tax Credits, p. 10.

[30] Two examples are aerospace and semiconductor chip manufacturers. See McGee Grisby and John Westmoreland, "The Research Tax Credit: A Temporary and Incremental Dinosaur," Tax Notes, vol. 93, no. 12, December 17, 2001, p. 1633.

[31] See CRS Report 95-871, Tax Preferences for Research and Experimentation: Are Changes Needed? by William A. Cox. (This report is out of print. Copies may be obtained from Gary Guenther at (202) 707-7742, upon request.) (Hereinafter cited as Cox, Tax Preferences for Research and Experimentation.)

[32] Ibid., p. 8.

[33] See, for example, Edwin Mansfield, The Positive Sum Strategy, pp. 309-311.

[34] Ibid., p. 9.

[35] In the case of capital-intensive projects, 50% of outlays go to structures and equipment, 35% qualify for expensing and the credit, and 15% qualify for expensing alone. In the case of intermediate projects, 30% of outlays go to structures and equipment, 50% qualify for expensing and the credit, and 20% qualify for expending alone. And in the case of labor-intensive projects, 15% of outlays go to structures and equipment, 65% qualify for expensing and the credit, and 20% qualify for expensing only.

[36] See Jane G. Gravelle, "Effects of the 1981 Depreciation Revisions on the Taxation of Income from Business Capital," National Tax Journal, vol. 35, no. 1, March 1982, pp. 2-3.

[37] Cox, Tax Preferences for Research and Experimentation, p. 15.

[38] Ibid., p. 17.

[39] GAO, The Research Tax Credit's Design and Administration Can Be Improved, p. 14.

[40] Ibid., p. 27.

[41] Ibid., p. 27.

[42] For further discussion of the possible benefits to small firms of making the credit wholly or partially refundable, see Scott J. Wallsten, "Rethinking the Small Business Innovation Research Program," in Investing in Innovation, Lewis M. Branscomb and James H. Keller, eds. (Cambridge, MA: MIT Press, 1998), pp. 212-214.

[43] Michael D. Rashkin, "The Dysfunctional Research Credit Hampers Innovation," Tax Notes, June 6, 2011, p. 1066.

[44] Disputes over the credit are a recurring issue in IRS audits of large corporations in manufacturing, energy, chemicals, and information technology. See Alex E. Sadler and Jennifer A. Ray, "Navigating the Research Credit," Tax Notes, September 19, 2011, p. 1254.

[45] See P.L. 99-514, Section 231.

[46] U.S. Congress, Joint Committee on Taxation, General Explanation of the Tax Reform Act of 1986, JCS-10-87 (Washington: GPO, 1987), pp. 132-134.

[47] Sheryl Stratton and Barton Massey, "Major Changes to Research Credit Rules Sought at IRS Reg Hearing," Tax Notes, May 3, 1999, pp. 623-624.

[48] David L. Click, "Treasury Discovers Problems With New Research Tax Credit Regulations," Tax Notes, March 12, 2001, p. 1531.

[49] Sheryl Stratton, "Treasury Puts Brakes on Research Credit Regs; Practitioners Applaud," Tax Notes, vol. 90, no. 6, February 5, 2001, pp. 713-715.

[50] For more details on the latest set of proposed regulations and reactions to them in the business community, see David Lupi-Sher and Sheryl Stratton, "Practitioners Welcome New Proposed Research Credit Regulations," Tax Notes, December 24, 2001, vol. 93, no. 13, pp. 1662-1665.

[51] Alison Bennett, "IRS Issues Final Research Credit Rules With Safe Harbor For Qualified Activities," Daily Report for Executives, Bureau of National Affairs, December 23, 2003, p. GG-2.

[52] Christopher J. Ohmes, David S. Hudson, and Monique J. Migneault, "Final Research Credit Regulations Expected to Immediately Affect IRS Examinations," Tax Notes, February 23, 2004, p. 1024.

[53] Michael D. Rashkin, Research and Development Tax Incentives: Federal, State, and Foreign (Chicago: CCH Inc., 2003), p. 87.

[54] David L. Click, "Zeal and Activity in the Arena of the Research Tax Credit," Tax Notes, December 15, 2008, p. 1307.

[55] David Click, "Substantiating the Research Credit," The Tax Adviser, vol. 41, no. 4, April 2010, p. 227.

[56] GAO, The Research Tax Credit's Design and Administration Can Be Improved, p. 33.

[57] Rashkin, "The Dysfunctional Research Credit Hampers Innovation," pp. 1066-1067.

[58] Rashkin, The Dysfunctional Research Credit Hampers Innovation, p. 1069.

[59] Joint Committee on Taxation, Estimated Revenue Effects of H.R. 4438, JCX-50-14 (Washington: May 8, 2014).

[60] Meg Shreve, "White House Vows Veto of House Research Credit Bill," Tax Notes, May 12, 2014.

[61] Department of the Treasury, General Explanations of the Administration's Fiscal Year 2015 Revenue Proposals (Washington: March 2014), pp. 12 and 280.

[62] Joint Committee on Taxation, Estimated Revenue Effects of the "Tax Reform Act of 2014," JCX-20-14 (Washington: February 26, 2014), p. 6.

In: Federal Influence on Innovation
Editor: Rachael A. Young
ISBN: 978-1-63463-627-8

Chapter 3

THE ROLE OF TRADE SECRETS IN INNOVATION POLICY*

John R. Thomas

SUMMARY

Many businesses have developed proprietary information that provides a competitive advantage because it is not known to others. As the United States continues its shift to a knowledge- and service-based economy, the strength and competitiveness of domestic firms increasingly depends upon their know-how and intangible assets. Trade secrets are the form of intellectual property that protects this sort of confidential information.

Trade secret law protects secret, valuable business information from misappropriation by others. Subject matter ranging from marketing data to manufacturing know-how may be protected under the trade secret laws. Trade secret status is not limited to a fixed number of years, but endures so long as the information is valuable and maintained as a secret. A trade secret is misappropriated when it has been obtained through the abuse of a confidential relationship or improper means of acquisition.

A number of competing innovation policy concerns help shape the particular doctrines that comprise trade secret law. The availability of legal protection for trade secrets potentially promotes innovation,

* This is an edited, reformatted and augmented version of a Congressional Research Service publication R41391, prepared for Members and Committees of Congress dated January 15, 2014.

encourages firms to invest in employee development, and confirms standards of commercial ethics and morality. On the other hand, trade secret protection involves the suppression of information, which may hinder competition and the proper functioning of the marketplace. An overly robust trade secret law also could restrain employee mobility and promote investment in costly, but socially inefficient security measures.

Trade secrets are primarily a matter of state law. In 1996, Congress enacted the Economic Espionage Act (EEA), a statute that criminalizes both "economic espionage" and the "theft of trade secrets." The EEA provides for substantial fines and imprisonment penalties, as well as criminal forfeiture of property and court orders preserving the confidentiality of trade secrets. Some commentators believe that few prosecutions have occurred under the EEA since its enactment and have deemed the legislation ineffective.

Patents and trade secrets provide different intellectual property options for many new inventions. Inventors typically must choose (1) to maintain an invention as a trade secret, (2) to obtain a patent on the invention, or (3) to allow the invention to enter the public domain. As a result, federal legislation or other developments that are perceived to alter the effectiveness of the patent system may make the trade secret more or less attractive to industry.

As of the date of publication of this report, two bills have been introduced in the 113th Congress that focus upon trade secrets. H.R. 2466, the Private Right of Action Against Theft of Trade Secrets Act of 2013, would introduce a private cause of action within the EEA. S. 1770, the Future of American Innovation and Research Act of 2013, would establish a private cause of action against a person who misappropriates a trade secret while located outside the United States, provided that the misappropriation causes or is reasonably anticipated to cause an injury within the United States or to a U.S. person. As of the date of publication of this report, neither bill had been enacted.

INTRODUCTION

In recent years Congress has shown considerable interest in promoting the strength and international competitiveness of U.S. industry. Several recent legislative proposals have concerned the intellectual property laws, a widely recognized mechanism for promoting innovation. In the 113th Congress, H.R. 2466, the Private Right of Action Against Theft of Trade Secrets Act of 2013, would introduce a private cause of action for misappropriation of trade secrets under federal law. S. 1770, the Future of American Innovation and Research Act of 2013, would establish a private cause of action against a person who misappropriates a trade secret while located outside the United States, provided that the misappropriation causes or is reasonably anticipated to cause an injury within the United States or to a U.S. person. As of the date of publication of this report, neither bill had been enacted.

The term "trade secret" generally refers to secret, commercially valuable information, including such subject matter as confidential formulae, techniques for manufacturing a product, and customer lists.[1] Trade secret protection is largely a matter of state law.[2] Under those laws, misappropriation of a trade secret may be enjoined by a court and the defendant may also be liable for compensatory and punitive damages.[3] One notable federal statute, the Economic Espionage Act of 1996,[4] makes the theft or misappropriation of a trade secret a federal crime under certain circumstances.

Trade secrets play a role in U.S. innovation policy. Trade secrets may establish incentives to innovate because they provide a mechanism for firms to capture the benefits of their inventions.[5] Yet trade secrets also have proven controversial because they suppress, rather than disclose, particular innovations to the public.[6] They also may have a significant impact upon the mobility of highly trained employees between firms.[7] Further, because innovators often face a mutually exclusive choice between patenting their inventions or maintaining them as trade secrets,[8] alterations to one of these regimes may alter the perceived attractiveness of the other.[9]

This report provides an overview of the law and policy of trade secrets. It discusses the role of trade secrets in U.S. innovation policy. It then reviews the sources of trade secret law and the substantive rules that they provide. The report then provides a more detailed review of existing federal legislation that pertains to trade secrets. In its next section, the report then discusses the relationship between patent law and trade secret law. The report closes with an identification of congressional issues and options within this field.

TRADE SECRETS AND INNOVATION POLICY

Many businesses have developed proprietary information that provides a competitive advantage because it is not known to others.[10] This category may include "high-tech" information such as chemical formulae, manufacturing techniques, product design, and technical data. But it may also include relatively "low-tech" information such as customer lists, business leads, marketing strategies, pricing schedules, and sales techniques.[11] Potentially of additional competitive value is "negative know-how": previously attempted, but flawed techniques or "blind alleys" that did not achieve their intended results.[12]

As the United States continues its shift to a knowledge- and service-based economy, the economic strength and competitiveness of firms increasingly depend upon their know-how and intangible assets. Leonard I. Nakamura, Assistant Vice President and Economist of the Federal Reserve Bank of Philadelphia, explains that in the U.S economy over the past half-century, "mass production and tangible investment have become less important, while new products ... and intangible investment have become more important."[13] One recent estimate placed the value of trade secrets owned by U.S. publicly traded companies at five trillion dollars.[14]

Another way of measuring the increasing importance of intangible assets is by considering the value of the Standard & Poor's 500. The "S&P 500" consists of the marketplace value of 500 large publicly held companies.[15] In 1975, 16.8% of the total value of the S&P 500 reportedly consisted of intangible assets. In 2005, intangible assets reportedly constituted 79.7% of the total value of these firms.[16] According to attorney R. Mark Halligan, "the vast bulk of intangible assets are trade secret assets."[17]

Yet the rise of computer technology, the ubiquity of cell phones and the Internet, and our transition to the Information Age have increased the difficulty that firms encounter in maintaining the confidentiality of their proprietary information. Years ago, the theft of trade secrets may have involved the taking of laboratory notebooks, memoranda, or other papers from a competitor's office despite the presence of security personnel or surveillance cameras. Today, a trade secret misappropriator can download proprietary information from company computers, or take photographs of confidential documents using a cell phone camera, within moments. As attorney Victoria A. Cundiff concisely states: "The digital world is no friend to trade secrets."[18]

As U.S. firms become increasingly immersed in global competition, some observers believe that foreign firms and even foreign governments have devoted significant resources towards industrial espionage.[19] According to the Office of the National Counterintelligence Executive, "[t]he United States remains the prime target for foreign economic collection and industrial espionage by virtue of its global technological leadership and innovation."[20] These efforts have been linked not just to the economic competitiveness of the United States, but also to its national security.[21]

The legal concept of trade secrets addresses these circumstances. Principles of trade secret law guard against the misappropriation of information that is not generally known. In view of the increasing prominence of information, it is unsurprising that Michael Risch, a member of the law faculty of the West Virginia University College of Law, asserts that trade secrets are "arguably the most important and most heavily litigated intellectual property right."[22]

Framing the trade secret law requires a balancing of competing interests. A robust trade secret law that provides strong protection to proprietary commercial information potentially holds many advantages. It may allow firms to capture the benefits of the costs and time it takes to develop the information, without having to share the benefits of that information with others. Trade secret law therefore may be seen as providing incentives to innovate.[23]

Trade secret law may also encourage firms to invest in human capital. A firm is more likely to invest in employee development if it has some confidence that employee cannot immediately use his knowledge in the service of a competitor.[24] Firms may also establish a trade secret easily through self-help measures. Commercially valuable information is protected once a firm makes reasonable efforts to maintain it in confidence. There is no need for formal government involvement, in contrast to patents.[25]

Trade secret law also confirms and regulates standards of commercial ethics and morality.[26] The misappropriation doctrine applies only against wrongdoers—those who have breached a duty of confidence, engaged in espionage, or otherwise acted in bad faith.[27] Trade secret law thus recognizes that even within a marketplace based open free competition, certain kinds of competitive behavior step beyond our social norms and should be discouraged.[28]

On the other hand, trade secret laws potentially have negative aspects. First, although trade secret law may promote advancement, it might facilitate a particular kind of innovation—the development of information that is itself amenable to being kept secret.[29] In addition, the protection of trade secrets necessarily requires firms to conceal their new developments. But as Judge Goldberg observed 40 years ago, "for our industrial competition to remain healthy there must be breathing room for observing a competing industrialist."[30] Thus, while some degree of information protection may be needed to promote innovation, some amount of information sharing may be essential for competition and proper functioning of the market.[31]

As well, firms must expend resources to maintain information as a trade secret. Employees must sign confidentiality agreements, locks and safes must be installed, and electronic protection measures must be in place on computer systems. These measures entail time and expense.[32] Employers also may be encouraged to limit access to valuable information to select employees. Disclosing valuable trade secrets on a "need to know" basis to one's employees may restrict employee development and ultimately hinder the operation of the firm.[33]

Trade secret law may negatively impact employee mobility. Individuals who are unable to take their knowledge from job to job may be limited in their ability to change employers.[34] As explained by Alan L. Durham, a member of the law faculty of the University of Alabama, an overly robust view of trade secret law potentially may "limit individual freedom, weaken employee bargaining power, and harm society through diminished competition."[35]

The various rules that together comprise the discipline of trade secret law may be fashioned in an effort to maximize the potential advantages of trade secrets while minimizing their disadvantages. This report next provides a more detailed review of the doctrines that regulate the acquisition and enforcement of trade secrets.

AN OVERVIEW OF TRADE SECRET LAW

Basic Principles

While it has been written that an "exact definition of a trade secret is not possible,"[36] a trade secret generally consists of secret, commercially valuable information.[37] As explained by Henry Perritt, Jr., a member of the faculty of the Chicago-Kent College of Law, "trade secret subject matter includes any of the major functions of business enterprise: production and operations, engineering and research and development, marketing, finance, purchasing, and management."[38] Specific examples of trade secret subject matter include customer lists, manufacturing processes, marketing strategies, pricing information, product design, recipes, and sales techniques.[39] One court has described trade secrets as follows:

> A trade secret is really just a piece of information (such as a customer list, or a method of production, or a secret formula for a soft drink) that the holder tries to keep secret by executing confidentiality agreements with employees and others and by hiding the information from outsiders by means of fences, safes, encryption, and other means of concealment, so that the only way the secret can be unmasked is by a breach of contract or a tort.[40]

Whether information qualifies as a "trade secret" is a question of fact that may be determined by a jury. Among the factors in assessing whether certain subject matter is a trade secret are:

- the extent to which the information is known outside of the company;
- the extent to which it is known by employees and others involved in the company;
- the extent of measures taken by the company to guard the secrecy of the information;
- the value of the information to the company and to its competitors;
- the amount of effort or money expended by the company in developing the information; and
- the ease or difficulty with which the information could be properly acquired or duplicated by others.[41]

The law protects trade secrets from misappropriation by others. Misappropriation is a tort that may occur in several distinct ways. One is when an individual acquires the trade secret through improper means, such as theft, bribery, misrepresentation, or espionage.[42] Another is when the individual uses or discloses the trade secret through a breach of confidence. For example, an employee might switch jobs, and then disclose his previous employer's trade secrets in violation of a confidentiality agreement.[43]

Finally, a trade secret may be misappropriated if it is used or disclosed with knowledge that the trade secret had been acquired improperly or through mistake. A person who uses information that he knows to have been stolen by another is therefore also guilty of misappropriation.[44] Misappropriation of a trade secret may be enjoined by a court and the defendant may also be liable for compensatory and punitive damages.[45]

Conversely, it is not a violation of trade secret law for another firm to discover the subject matter of a trade secret independently.[46] "Reverse engineering" is also considered to be an appropriate means for one firm to acquire the subject matter of another's trade secret.[47] A firm that discerns the subject matter of the trade secret by inspecting products available to the public also has not engaged in misappropriation.[48]

Trade secret protection may extend indefinitely. So long as information is not generally known to the public, confers an economic benefit to its holder, and is subject to reasonable efforts to maintain its secrecy, it may be considered a trade secret.[49] However, the trade secret status of information may be lost if the information is accidentally or intentionally disclosed by an employee.[50] Once a trade secret has been exposed to the public, its protected character is lost and cannot later be retrieved.[51] However, disclosures of trade secrets to third parties for certain limited reasons do not waive trade secret protections, so long as the trade secret owner took reasonable measures to maintain its secrecy before and during disclosure, such as requiring non-disclosure or confidentiality agreements from each recipient of confidential information.[52]

Sources of Law

The discipline of trade secrets was traditionally developed through state common law. State courts developed the essential principles of trade secret law through their decisions, which were then observed as precedent in subsequent litigation. In 1939, the American Law Institute (ALI), a group of lawyers, judges, and legal scholars, published a treatise titled the "Restatement of Torts." The Restatement of Torts included two sections dealing with the law of trade secrets. Section 757 explained the subject matter of trade secrets, while Section 758 spelled out the elements of a trade secret misappropriation cause of action. Although this treatment was succinct, many commentators believe that these definitions proved influential in the courts.[53]

Trade secrets were not addressed in the Second Restatement of Torts published in 1978. The ALI at that time concluded that trade secret law had grown "no more dependent on Tort law than it is on many other general fields of law and upon broad statutory developments,"[54] and opted not to house trade secrets there. The ALI addressed this gap in its 1993 Restatement (Third) of Unfair Competition, which deals with trade secrets in sections 39–45.

In addition, the National Conference of Commissioners on Uniform State Law (NCCUSL) issued the Uniform Trade Secrets Act (UTSA) in 1979. The NCCUSL consists of a group of academics, attorneys, and judges who draft statutes addressing a variety of issues, and then propose that each state enact them.[55] The NCCUSL lacks direct legislative authority itself. Its uniform acts become law only to the extent that state legislatures adopt them.

The UTSA has arguably proven successful, as it has reportedly been enacted in 47 states, the District of Columbia, and the Virgin Islands.[56] The three states that have reportedly not enacted the UTSA are Massachusetts, North Carolina, and New York. These states do recognize trade secrets, but provide protection through a distinct statute or the common law.

It should be appreciated that many jurisdictions have enacted the UTSA after making some changes to the original text of the proposed legislation. Opinions vary on how significant these modifications have been. While some commentators view state-by-state variations as "generally minor,"[57] others have opined that "they include fundamental differences about what constitutes a trade secret, what is required to misappropriate it, and what remedies are available."[58]

THE ECONOMIC ESPIONAGE ACT

As noted previously, trade secrets have traditionally been the subject of state law. Prior to 1996, arguably the most significant federal legislation on point was the Trade Secrets Act.[59] Although broadly titled, this 1948 statute is actually of narrow application. It forbids federal government employees and government contractors from making an unauthorized disclosure of confidential government information, including trade secrets. The sanctions for violating this criminal offense are removal from office or employment, and a fine and/or imprisonment of not more than one year. The law does not apply to state or local government actors or to private sector employees.

In 1996, motivated by concerns over growing international and domestic economic espionage against U.S. firms, Congress enacted new legislation pertaining to trade secrets. The Economic Espionage Act (EEA) criminalizes both "economic espionage" and the "theft of trade secrets." The "economic espionage" provision punishes those who knowingly misappropriate, or attempt or conspire to misappropriate, trade secrets with the intent or knowledge that the offense will benefit a foreign government, instrumentality or agent.[60] The "theft of trade secrets" prohibition is of more general application. The principal elements of an EEA claim for theft of trade secrets are:

1. the intentional and/or knowing theft, appropriation, destruction, alteration, or duplication of
2. a trade secret placed in interstate commerce (3) with intent to convert the trade secret and (4) intent or knowledge that such action will injure the owner.[61]

The EEA provides for substantial fines and imprisonment penalties. For economic espionage, the maximum penalties increase to $500,000 for individuals and imprisonment of 15 years, or $10 million for corporations.[62] Theft of trade secrets is punishable by a fine of up to $250,000 for individuals as well as imprisonment of up to 10 years. Organizations can be fined up to $5 million.[63] The EEA also provides for criminal forfeiture of property and court orders preserving the confidentiality of trade secrets.[64]

The EEA has been subject to critical commentary. Attorney R. Mark Halligan expressed the view that the legislation was "ineffective" and observed, in 2008, that there had been fewer than 60 prosecutions in keeping with its provisions.[65] Susan W. Brenner, a member of the law faculty of the University of Dayton, and security expert Anthony C. Crescenzi opine that the "paucity" of prosecutions under the EEA are due to a number of factors, including the complexity of the cases, the desire of the Department of Justice only to bring cases it can win, the diplomatic repercussions of bringing such a case, and the unwelcome possibility of additional disclosure of trade secrets during the litigation.[66] Brenner and Crescenzi conclude that the "individual and combined effect of the systemic factors discussed above is to erode the EEA's effectiveness as a weapon against economic espionage."[67]

TRADE SECRETS AND PATENTS

Trade secrets and patents form two distinct fields within the U.S. intellectual property system. In one sense, trade secrecy serves as the chief alternative to the patent system. Most inventors must choose one of three options: (1) maintain a technology as a trade secret, (2) seek patent protection, (3) or decline to seek intellectual property protection at all and allow the technology to enter the public domain.[68] In view of the close relationship between trade secrets and patents, this report next provides an overview of the patent system and its interaction with trade secret law.

Introduction to the Patent System

An inventor may seek the grant of a patent by preparing and submitting an application to the U.S. Patent and Trademark Office, or USPTO.[69] USPTO officials known as examiners then determine whether the invention disclosed in the application merits the award of a patent.[70] The USPTO examiner will consider a number of legal requirements, including whether the submitted application fully explains and distinctly claims the invention. In particular, the application must enable persons skilled in the art to make and use the invention without undue experimentation. In addition, the application must provide the "best mode," or preferred way, that the applicant knows to practice the invention.[71]

The examiner will also determine whether the invention itself fulfills certain substantive standards set by the patent statute. To be patentable, an invention must meet four primary requirements. First, the invention must fall within at least one category of patentable subject matter.[72] According to the Patent Act, an invention which is a "process, machine, manufacture, or composition of matter" is eligible for patenting. Second, the invention must be useful, a requirement that is satisfied if the invention is operable and provides a tangible benefit.[73] Third, the invention must be novel, or different, from subject matter disclosed by an earlier patent, publication, or other state-of-the-art knowledge.[74] Finally, an invention is not patentable if "the subject matter as a whole would have been obvious at the time the invention was made to a person having ordinary skill in the art to which said subject matter pertains."[75] This requirement of "nonobviousness" prevents the issuance of patents claiming subject matter that a skilled artisan would have been able to implement in view of the knowledge of the state of the art. If the USPTO allows the patent to issue, its owner obtains the right to exclude others from making, using, selling, offering to sell, or importing into the United States the patented invention.[76]

Trade Secrets and Patents Compared

Inventors who do not wish to dedicate their technologies to the public domain must, as a general matter, choose between trade secret and patent protection. A number of factors inform this decision. One is whether the inventor will practically be able to keep the technology secret. A knowledgeable observer may readily be able to inspect a motor, machine, or other mechanical technology in order to learn its design, for example. On the other hand, the composition of a chemical compound may be much more difficult to discern.[77]

The costs associated with acquiring and maintaining patents are another factor. In this respect, it should be appreciated that a U.S. patent provides rights only with the United States. However, virtually anyone in the world may review a U.S. patent to learn of its contents. As a result, U.S. inventors may need to obtain patents in many foreign countries in order to secure meaningful protection.

In addition, the process of acquiring a patent may take many years.[78] A USPTO examiner in 2009 would not review a patent application until, on average, 25.8 months after it was filed.[79] The "first action pendency" during 2000 was 13.6 months.[80] Many observers believe that if current conditions continue, the backlog and delay are likely to grow at the USPTO in coming years.[81] These delays may prove too lengthy for innovators in fast-moving industries, suggesting that trade secret protection is the more appropriate choice.

Also, trade secrets may potentially extend indefinitely, so long as the requirements for trade secret protection are maintained. In contrast, patents expire after a set period of time, normally 20 years after the date they were filed. On the other hand, trade secret protection may be lost through a competitor's reverse engineering or independent discovery. As explained by Dan Burk, a member of the faculty of the University of California, Irvine School of Law, "the inventor's choice is an election between twenty years of certain patent protection or perpetual, but less certain, trade secret protection."[82]

It should also be appreciated that when an inventor obtains a patent on an invention, the USPTO publishes that patent in a formal document. That publication destroys the trade secret status of any previously confidential information disclosed within it.[83] In addition, the USPTO publishes many, but not all, pending patent applications "promptly after the expiration of a period of 18 months" after they are filed.[84] This measure also destroys the trade secret status of information contained within the published application, even if the USPTO subsequently rejects the application and no patent ever issues on that invention.[85]

The Prior Commercial User Defense

The trade secret and patent systems are sometimes viewed as acting in conflicting ways. Trade secret protection is predicated upon the maintenance of the protected information in confidence. In contrast, each patented invention is the subject of a formal document, the patent instrument, which provides a complete description of the invention. As a result, while the patent system appears to promote the public disclosure of new technologies, the trade secret discourages disclosure.[86] As described previously, it could be argued that trade secret law encourages the development of technologies that are capable of being kept secret.[87]

However, some commentators believe that patents and trade secrets generally act in a complementary manner.[88] Mark Lemley, a member of the faculty of the Stanford Law School, has explained that trade secret law provides valuable incentives to innovate in areas where the patent law does not reach, such as customer lists and business plans.[89] Lemley further explains that although the law requires that reasonable efforts must be made to maintain secrecy, absent trade secret law, firms might need to engage in even more physical and contractual measures to prevent disclosure. As a result, a society without trade secret law might potentially have more, rather than less secrecy.[90]

In any event, Congress modified the usual rules governing the relationship between trade secrets and patents when it established the "first inventor defense" in the American Inventors Protection Act of 1999,[91] which in turn was replaced by the "prior commercial user defense" in the Leahy-Smith America Invents Act (AIA) in 2011.[92] The earlier, 1999 legislation in part provided an infringement defense for an earlier inventor of a "method of doing or conducting business" that was later patented by another. The AIA subsequently expanded the range of individuals who may assert this defense in court. Even more significantly, the new legislation eliminated the restriction of the prior commercial use defense to business method patents. Under the AIA, a patent claiming any type of invention may be subject to the prior commercial user defense.

The prior commercial user defense accounts for the complex relationship between the law of trade secrets and the patent system. Trade secrecy protects individuals from misappropriation of valuable information that is useful in commerce. One reason an inventor might maintain the invention as a trade secret rather than seek patent protection is that the subject matter of the invention may not be regarded as patentable. Such inventions as customer lists or data compilations have traditionally been regarded as amenable to trade secret protection but not to patenting.[93] Inventors might also maintain trade secret protection due to ignorance of the patent system or because they believe they can keep their invention as a secret longer than the period of exclusivity granted through the patent system.[94]

The patent law does not favor trade secret holders, however. Well-established patent law provides that an inventor who makes a secret, commercial use of an invention for more than one year prior to filing a patent application at the USPTO forfeits his own right to a patent.[95] This policy is based principally upon the desire to maintain the integrity of the statutorily prescribed patent term. The patent law grants patents a term of 20 years, commencing from the date a patent application is filed.[96] If the trade secret holder could make commercial use of an invention for many years before choosing to file a patent application, he could disrupt this regime by delaying the expiration date of his patent.[97]

Settled patent law principles established that prior secret uses would not defeat the patents of later inventors.[98] If an earlier inventor made secret commercial use of an invention, and another person independently invented the same technology later and obtained patent protection, then the trade secret holder could face liability for patent infringement. This policy is based upon the reasoning that once issued, published patent instruments fully inform the public about the invention, while trade secrets do not. Between a subsequent inventor who patented the invention, and thus had disclosed the invention to the public, and an earlier trade secret holder who had not, the law favored the patent holder.

An example may clarify this rather complex legal situation. Suppose that Inventor A develops and makes commercial use of a new manufacturing process. Inventor A chooses not to obtain patent protection, but rather maintains that process as a trade secret. Many years later, Inventor B independently develops the same manufacturing process and promptly files a patent application claiming that invention. In such circumstances, Inventor A's earlier, trade secret use does not prevent Inventor B from procuring a patent. Furthermore, if the USPTO approves the patent application, then Inventor A faces infringement liability should Inventor B file suit against him.

The American Inventors Protection Act of 1999 somewhat modified this principle. That statute in part provided an infringement defense for an earlier user of a "method of doing or conducting business" that was later patented by another. By limiting this defense to patented methods of doing business, Congress responded to the 1998 Federal Circuit opinion in *State Street Bank and Trust Co. v. Signature Financial Group.*[99] That judicial opinion recognized that business methods could be subject to patenting, potentially exposing individuals who had maintained business methods as trade secrets to liability for patent infringement.

Again, an example may aid understanding of the prior commercial user defense. Suppose that Inventor X develops and exploits commercially a new method of doing business. Inventor X maintains his business method as a trade secret. Many years later, Inventor Y independently develops the same business method and promptly files a patent application claiming that invention. Even following the enactment of the American Inventors Protection Act, Inventor X's earlier, trade secret use would not prevent Inventor Y from procuring a patent. However, should the USPTO approve Inventor Y's patent application, and should Inventor Y sue Inventor X for patent infringement, then Inventor X may potentially claim the benefit of the first inventor defense. If successful,[100] Inventor X would enjoy a complete defense to infringement of Inventor Y's patent.

Prior to the AIA, the prior commercial user defense could "be asserted only by the person who performed the acts necessary to establish the defense.... "[101] The AIA also allowed the defense to be asserted by "any other entity that controls, is controlled by, or is under common control with such person.... "[102]

In addition, the AIA eliminated the restriction of the prior commercial user defense to business method patents. As a result, any type of patented invention may be subject to the prior commercial user defense. The new legislation also exempted patents held by universities from the prior commercial user defense when it stipulates that this is not available if "the claimed invention ... was made, owned or subject to an obligation of assignment to either an institution of higher education ... or a technology transfer organization whose primary purpose is to facilitate the commercialization of technologies developed by one or more such institutions of higher education."

Congressional Issues and Options

A variety of options are available for Congress with respect to trade secrets. If the current situation is deemed appropriate, then no action need be taken. Alternatively, Congress may wish to consider the adoption of a federal trade secret law. Several commentators believe that this step would promote the uniformity of trade secret law throughout the United States.[103] As attorney David S. Almeling asserts:

> Trade secrets stand alone as the only major type of intellectual property governed primarily by state law. Trademarks, copyrights, and patents are each governed primarily by federal statutes. Trade secrets, by contrast, are governed by fifty state statutes and common laws. The result is that trade secret law differs from state to state. It is time to eliminate

> these differences—and the significant problems they cause—by enacting a Federal Trade Secrets Act ("FTSA").[104]

According to Alemeling, general federal trade secret legislation would establish greater uniformity in substantive and procedural law than is possible in a state-based regime.

Other commentators have further suggested that the current state-based trade secrets system places the United States in violation of its obligations under two international agreements: the North American Free Trade Agreement (NAFTA) and the World Trade Organization Agreement on Trade-Related Aspects of Intellectual Property (TRIPS Agreement).[105] Both NAFTA and the TRIPS Agreement require member states to provide certain levels of trade secret protection. Because the portions of NAFTA and the TRIPS Agreement concerning trade secrets were modeled after the Uniform Trade Secrets Act (UTSA), those states that have adopted the UTSA without restrictive modifications likely comply with these international standards. But some states have not adopted the UTSA, and some states that have done so have arguably included more restrictive standards. Some commentators have asserted that these states place the United States in violation of NAFTA and the TRIPS Agreement. However, others have observed that any shortcomings of U.S. law on this point have yet to be challenged under either international agreement.[106]

Still other observers assert that no compelling case has been made to federalize trade secret law. Some believe that this step might create additional burdens and costs upon the federal judiciary.[107] Others cite federalism concerns, believing that the states possess a strong interest in regulating local economies in view of their own, local norms.[108] In addition, variation between the laws of the different states does not necessarily compel federalization of the field. For example, meaningful distinctions between the states exist in other areas of law, including such fundamental disciplines as contract law. Yet these disciplines remain subject to state law.[109]

A second possibility for Congress is to amend the Economic Espionage Act so as to create a private, federal cause of action for trade secret misappropriation. H.R. 2466, the Private Right of Action Against Trade Secrets Act of 2013, would do just that. Under that proposed legislation, any person who suffers an injury from trade secret misappropriation "may maintain a civil action against the violator to obtain appropriate compensatory damages and injunctive relief or other equitable relief."[110] H.R. 2466 does not appear wholly to federalize trade secret law—under this proposal, trade secret law would remain primarily a matter of state law. This legislation had not been enacted as of the publication of this report.

S. 1770, the Future of American Innovation and Research Act of 2013, provides another possibility for future reform efforts. This bill would establish a private cause of action against a person who misappropriates a trade secret while located outside the United States, provided that the misappropriation causes or is reasonably anticipated to cause an injury within the United States or to a U.S. person. This legislation had not been enacted as of the publication of this report.

In addition, Congress may wish to remain apprised of the potential effect of continuing patent reform efforts upon trade secrets. As this report has discussed, trade secrets and patents act in complementary ways to protect innovation in the United States.[111] Further, to some degree the two forms of intellectual property act as imperfect substitutes for each other. As a result, legislative reforms that are perceived to make patents more effective may reduce industrial reliance upon trade secrets. Conversely, amendments to the Patent Act that are believed to reduce the effectiveness of patents may increase the willingness of firms to retain information as trade secrets.[112]

CONCLUDING OBSERVATIONS

Trade secrets form a significant component of the intellectual property system of the United States. The importance of trade secrets will likely increase as U.S. industry continues to participate within a knowledge-based, global economy with increasingly sophisticated competitors. Because trade secrets are currently a matter of state law, congressional influence over the system has thus far been indirect: Through the enactment of a criminal statute, the Economic Espionage Act, and through amendment to the patent law. Whether further intervention is required in the U.S. trade secret system remains a matter of congressional judgment.

End Notes

[1] See Uniform Trade Secret Act, §1.

[2] See Michael Traynor and Katy Hutchinson, "Some Open Questions About Intellectual Property Remedies," 14 *Lewis & Clark Law Review* (2010), 453.

[3] See Shubha Ghosh, "Open Borders, Intellectual Property & Federal Criminal Trade Secret Law," 9 *John Marshall Review of Intellectual Property Law* (Fall 2009), 24.

[4] 104th Congress, P.L. 104-294 (Oct. 11, 1996).

[5] See Michael Risch, "Trade Secret Law and Information Development Incentives," in *The Law and Theory of Trade Secrecy: A Handbook of Contemporary Research* (Rochelle C. Dreyfuss, Katherine J. Strandburg, eds., Edward Elgar Publishing, 2010).

[6] See Michael P. Simpson, "The Future of Innovation: Trade Secrets, Property Rights, and Protectionism—An Age-Old Tale," 70 *Brooklyn Law Review* (2005), 1121.

[7] See Charles Tait Graves, "Trade Secrets as Property: Theory and Consequences," 15 *Journal of Intellectual Property Law* (2007), 39.

[8] See Daniel C. Munson, "The Patent-Trade Secret Decision: An Industrial Perspective," 78 *Journal of the Patent and Trademark Office Society* (1996), 689.

[9] See Douglas Lichtman, "How the Law Responds to Self-Help," 1 *Journal of Law, Economics & Policy* (Winter 2005), 215 (observing that trade secret law acts "as both a complement to and competitor for patent law.").

[10] See Kurt M. Saunders, "The Law and Ethics of Trade Secrets: A Case Study," 42 *California Western Law Review* (2006), 209.

[11] Milton Babirak, *Secrets of the Uniform Trade Secret Act*, D.C. BAR CLE PROGRAM (Nov. 9, 2005).

[12] Charles Tait Graves, "The Law of Negative Knowledge: A Critique," 15 *Texas Intellectual Property Law Journal* (2007), 387.

[13] Leonard I. Nakamura, *Intangible Assets and National Income Accounting: Measuring a Scientific Revolution*, Working Paper No. 09-11, Federal Reserve Bank of Philadelphia, at 3.

[14] See Elizabeth A. Rowe, "Contributory Negligence, Technology, and Trade Secrets," 17 *George Mason Law Review* (2009), 1.

[15] See http://www.standardandpoors.com.

[16] James E. Malackowski, "The Intellectual Property Marketplace: Past, Present and Future, 5 *The John Marshall Journal of Intellectual Property Law* (2006), 605.

[17] R. Mark Halligan, "Protection of U.S. Trade Secret Assets: Critical Amendments to the Economic Espionage Act of 1996," 7 *John Marshall Review of Intellectual Property Law* (2008), 656.

[18] Victoria A. Cundiff, "Reasonable Measures to Protect Trade Secrets in a Digital Environment," 49 *IDEA: The Intellectual Property Law Review* (2009), 359.

[19] Susan W. Brenner and Anthony C. Crescenzi, "State-Sponsored Crime: The Futility of the Economic Espionage Act," 28 *Houston Journal of International Law* (2006), 389.

[20] Office of the National Counterintelligence Executive, *Annual Report to Congress on Foreign Economic Collection and Industrial Espionage* (2007), 1.

[21] Aaron J. Burstein, "Trade Secrecy as an Instrument of National Security? Rethinking the Foundations of Economic Espionage," 41 *Arizona State Law Journal* (2009), 933.

[22] Michael Risch, "Why Do We Have Trade Secrets?," 11 *Marquette Intellectual Property Law Review* (2007), 1.

[23] Oren Bar-Gill, "Law and the Boundaries of Technology-Intensive Firms," 157 *University of Pennsylvania Law Review* (2009), 1649.

[24] Miles J. Feldman, "Toward a Clearer Standard of Protectable Information: Trade Secrets and the Employment Relationship," 9 *High Technology Law Journal* (1994), 151.

[25] Lemley, *supra,* at 313.

[26] See Elizabeth A. Rowe, "A Sociological Approach to Misappropriation," 58 *University of Kansas Law Review* (2009), 1.

[27] See Rebecca S. Eisenberg, "Proprietary Rights and the Norms of Science in Biotechnology Research," 97 *Yale Law Journal* (1987), 177.

[28] See James Grimmelmann, "The Ethical Visions of Copyright Law," 77 *Fordham Law Review* (2009), 2005.

[29] Michael P. Simpson, "The Future of Innovation: Trade Secrets, Property Rights, and Protectionism—An Age-Old Tale," 70 *Brooklyn Law Review* (2005), 1121.

[30] E.I. duPont deNemours & Co. v. Christopher, 431 F.2d 1012 (5th Cir. 1970).

[31] Michael J. Garrison and John T. Wendt, "The Evolving Law of Employee Noncompete Agreements: Recent Trends and an Alternative Policy Approach," 45 *American Business Law Journal* (2008), 107.

[32] See Douglas Lichtman, "How the Law Responds to Self-Help," 1 *Journal of Law, Economics & Policy* (2005), 215.

[33] See Bruce Fallick *et al.*, "Job-Hopping in Silicon Valley: Some Evidence Concerning the Microfoundations of a High-Technology Cluster," 88 *The Review of Economics and Statistics* (2006), 472.

[34] See Bruce Alan Kugler, "Limiting Trade Secret Protection," 22 *Valparaiso University Law Review* (1988), 725.

[35] Alan L. Durham, "Natural Laws and Inevitable Infringement," 93 *Minnesota Law Review* (2009), 933.

[36] Restatement (First) of Torts §757, comment b.

[37] Uniform Trade Secrets Act §1(4).

[38] Henry H. Perritt, Jr., *Trade Secrets: A Practitioner's Guide* §3:9 (2d ed. 2006).

[39] See Babirak, *supra*.

[40] ConFold Pac. v. Polaris Indus., 433 F.3d 952, 959 (7th Cir. 2006) (citations omitted).

[41] Restatement (First) of Torts §757, comment b.

[42] Restatement (Third) of Unfair Competition §40 (1994).

[43] See Jennifer Brockett, "Protecting Intellectual Property During Layoffs," 32 *Los Angeles Lawyer* (April 2009).

[44] Restatement (Third) of Unfair Competition §40 (1994).

[45] Restatement (Third) of Unfair Competition §§44, 45 (1994).

[46] Restatement (Third) of Unfair Competition §43 (1994).

[47] See Kristin L. Black, "Crimes of Fashion: Is Imitation Truly the Sincerest Form of Flattery?," 19 *Kansas Journal of Law and Public Policy* (2010), 515.

[48] Restatement (Third) of Unfair Competition §43 (1994).

[49] Perritt, *supra* , at §2:2.

[50] See Religious Tech. Ctr. v. Netcom On-Line Communication Servs., 923 F. Supp. 1231, 1256 (N.D. Cal. 1995).

[51] In re Remington Arms Co., 952 F.2d 1029, 1033 (8th Cir. 1991).

[52] 1 Roger Milgrim, *Milgrim on Trade Secrets* §1.04.

[53] See Katarzyna A. Czapracka, "Antitrust and Trade Secrets: The U.S. and EU Approach," 24 *Santa Clara Computer High Technology Law Journal* (2008), 207.

[54] ALI, Restatements of Torts (Second), Introduction (1979).

[55] For further information about the organization, see http://www.nccusl.org.

[56] See David S. Almeling, "A Statistical Analysis of Trade Secret Litigation in Federal Courts," 45 *Gonzaga Law Review* (2009-10), 291.

[57] Babirak, *supra*, at 3.

[58] Almeling, *Four Reasons*, *supra,* at 774.

[59] June 25, 1948, c. 645, 62 Stat. 79 (codified at 18 U.S.C. §1905).

[60] 18 U.S.C. §1831.

[61] 18 U.S.C. §1832.

[62] 18 U.S.C. §1831.

[63] 18 U.S.C. §1832.

[64] 18 U.S.C. §§1834, 1835.

[65] See Halligan, *supra*.

[66] Susan W. Brenner and Anthony C. Crescenzi, "State-Sponsored Crime: The Futility of the Economic Espionage Act," 28 *Houston Journal of International Law* (2006), 389.

[67] Ibid. at 439.

[68] See generally John S. Paniaguas and Craig William Mandell, "A Practitoner's Guide to Protecting Technology Assets," 20 *DePaul Journal of Art, Technology & Intellectual Property Law* (2010), 279.

[69] 35 U.S.C. §111.

[70] 35 U.S.C. §131.

[71] 35 U.S.C. §112.

[72] 35 U.S.C. §101.

[73] Ibid.

[74] 35 U.S.C. §102.

[75] 35 U.S.C. §103.

[76] 35 U.S.C. §271.

[77] See Daniel C. Munson, "The Patent-Trade Secret Decision: An Industrial Perspective," 78 *Journal of the Patent and Trademark Office Society* (1996), 689.

[78] See CRS Report R41261, *Deferred Examination of Patent Applications: Implications for Innovation Policy*, by John R. Thomas.

[79] USPTO, Performance and Accountability Report Fiscal Year 2009 (available at http://www.uspto.gov/web/offices/ com/annual/2009/mda_02_02.html).

[80] USPTO, 2003 Performance and Accountability Report (available at http://www.uspto.gov/about/stratplan/ar/2003/ 040201_patentperform.jsp).

[81] See, *e.g.*, Jon Dudas et al., "Let the PTO Pay Its Own Way," 198 New Jersey Law Journal no. 12 (Dec. 21, 2009), 975; Steven Andersen, "Out of Balance," Inside Counsel (Nov. 1, 2009).

[82] Dan L. Burk, "Legal Constraint of Genetic Use Restriction Technologies," 6 *Minnesota Journal of Law, Science, and Technology* (2004), 335.

[83] See BondPro Corp. v. Siemens Power Generation, Inc., 463 F.3d 703, 706 (7th Cir. 2006).

[84] 35 U.S.C. §122(b)(1).

[85] See BondPro v. Siemens, *supra*.

[86] Michael P. Simpson, "The Future of Innovation: Trade Secrets, Property Rights, and Protectionism—An Age-Old Tale," 70 *Brooklyn Law Review* (2005), 1121.

[87] Ibid.

[88] See, e.g., Karl F. Jorda, "Patent and Trade Secret Complementariness: An Unexpected Synergy," 48 *Washburn Law Journal* (2008), 1; Jonathan D. Carpenter, "Intellectual Property: The Overlap Between Utility Patents, Plant Patents, the PVPA, and Trade Secrets and the Limitations on that Overlap," 81 *North Dakota Law Review* (2005), 171.

[89] Mark A. Lemley, "The Surprising Virtues of Treating Trade Secrets as IP Rights," 61 *Stanford Law Review* (2008), 131.

[90] Ibid.

[91] The American Inventors Protection Act of 1999, P.L. 106-113, was part of the Intellectual Property and Communications Omnibus Reform Act of 1999, attached by reference to the

Consolidated Appropriations Act for Fiscal Year 2000. President Clinton signed this bill on November 29, 1999.

[92] P.L. 112-29.

[93] Restatement of Unfair Competition §39.

[94] David D. Friedman, "Some Economics of Trade Secret Law," 5 *Journal of Economic Perspectives* (1991), 61, 64.

[95] 35 U.S.C. §102(b). See Metallizing Engineering Co. v. Kenyon Bearing & Auto Parts, 153 F.2d 516 (2d Cir. 1946).

[96] 35 U.S.C. §154.

[97] The AIA states in part that an invention may not be patented if it was "in public use, on sale, or otherwise available to the public" anywhere in the world prior to the inventor's filing date. As noted, earlier case law had held that secret uses and sales offers could, in some circumstances, qualify as prior art as being a "public use" and "on sale" under §102(b) of the predecessor statute. Whether this law remains sound is subject to some debate. USPTO Examination Guidelines state that the "phrase 'public use' in AIA 35 U.S.C. § 102(a)(1) is treated as having the same meaning as 'in public use' in pre-AIA 35 U.S.C. § 102(b)." However, some legislative history of the AIA suggests that the rules might have changed. For example, during one debate Senator Kyl stated:

Public uses and sales of an invention will remain prior art, but only if they make the invention available to the public. An inventor's confidential sale of his invention, his demonstration of its use to a private group, or a third party's unrestricted but private use of the invention will no longer constitute prior art. Only the sale or offer for sale of the invention to the relevant public or its use in a way that makes it publicly accessible will constitute prior art.

See 157 *Cong. Rec.* S5320 (daily ed. September 6, 2011). Conclusive resolution of this issue may become the task of the judiciary or the subject of further legislation.

[98] W.L. Gore & Associates v. Garlock, Inc., 721 F.2d 1540 (Fed. Cir. 1983).

[99] 149 F.3d 1368 (Fed. Cir. 1998).

[100] As earlier codified at 35 U.S.C. §273, the prior commercial user defense is subject to a number of additional qualifications. First, the defendant must have reduced the infringing subject matter to practice at least one year before the effective filing date of the application. Second, the defendant must have commercially used the infringing subject matter prior to the effective filing date of the patent. Finally, any reduction to practice or use must have been made in good faith, without derivation from the patentee or persons in privity with the patentee.

[101] 35 U.S.C. §273(b)(6)

[102] P.L. 112-29 at §5.

[103] See, e.g., Marina Lao, "Federalizing Trade Secrets Law in an Information Economy," 59 *Ohio State Law Journal* 1633 (1998); Rebel J. Pace, "The Case for a Federal Trade Secrets Act," 8 *Harvard Journal of Law & Technology* (1995), 427.

[104] See Almeling, *supra*.

[105] See, e.g., Marina Lao, "Federalizing Trade Secrets Law in an Information Economy," 59 *Ohio State Law Journal* 1633 (1998); Rebel J. Pace, "The Case for a Federal Trade Secrets Act," 8 *Harvard Journal of Law & Technology* (1995), 427.

[106] See Almeling, *supra*.

[107] See, e.g., American Intellectual Property Law Association, *Report of the AIPLA Trade Secrets Committee* (2007) (available at http://www.aipla.org).

[108] Ibid.

[109] See generally Llewellyn Joseph Gibbons, "Stop Mucking Up Copyright Law: A Proposal for a Federal Common Law of Copyright," 35 *Rutgers Law Journal* (2004), 959.

[110] H.R. 2466 at §2.

[111] See also Scott W. Cummings, "The Role of Trade Secrets in Today's Nanotechnology Patent Environment," 5 *Nanotechnology Law & Business* (2008), 41.

[112] See Christopher A. Cotropia and Mark A. Lemley, "Copying in Patent Law," 87 *North Carolina Law Review* (2009), 1421 (providing a skeptical discussion of this point).

In: Federal Influence on Innovation
Editor: Rachael A. Young
ISBN: 978-1-63463-627-8

Chapter 4

TAILORING THE PATENT SYSTEM FOR SPECIFIC INDUSTRIES*

John R. Thomas

SUMMARY

Congressional interest in the patent system has been demonstrated by the enactment of the Leahy-Smith America Invents Act (AIA) in the 112th Congress. Most of the provisions of the AIA apply to any type of patented invention, whether it consists of a chemical compound, mechanical device, electrical circuit, or other technology. However, other AIA provisions are specific to particular types of inventions, including business methods, tax strategies, and human organisms. The AIA reflects the principle that, for the most part, patentable inventions are generally subject to the same statutory provisions. However, a number of exceptions exist to this concept of technological neutrality.

This blended architecture has for many years prompted inquiry into whether the patent system operates best as a uniform system that applies neutrally to all inventions, or whether it could or should be tailored to meet the specific needs of different industries. Technologies and industrial sectors arguably differ in ways salient to the patent system. Among these distinctions are the costs and risks of research and development, the availability of trade secret protection as an effective

* This is an edited, reformatted and augmented version of a Congressional Research Service publication R43264, prepared for Members and Committees of Congress, dated January 14, 2014.

alternative to patenting, the number of patents that cover a particular product, and the patterns of patent acquisition and enforcement of firms within that sector. The patent system involves a number of parameters that could potentially be adjusted to meet the needs of individual sectors, including the speed with which applications are reviewed, the scope of exclusive rights afforded by a patent, and the term of the patent.

While some observers suggest the desirability of sector-specific patent principles, others believe them to be infeasible and unwise. They observe that legislative efforts to define particular industries may prove difficult, that attorneys may sometimes be able to draft patents artfully so as to fall within a favored category, and that U.S. industry is dynamic and resistive to a static statutory definition. In addition, U.S. membership within the World Trade Organization (WTO) may limit the ability to tailor the patent system to account for different industries and inventions, to the extent that compliance with WTO standards is desired. The WTO-administered Agreement on Trade-Related Aspects of Intellectual Property, or TRIPS Agreement, in part requires WTO member states to make patent rights available without discrimination as to the field of technology. The TRIPS Agreement admits some exceptions exist to this principle of technological neutrality, however.

Should Congress believe current circumstances to be appropriate, then no action need be taken. To the degree WTO compliance is desired, Congress could also legislate along the lines permitted by the TRIPS Agreement. Notably, although the TRIPS Agreement generally disallows discrimination with respect to technological fields, it permits distinctions on other grounds. Congress could also make use of regulatory exclusivities and other complementary intellectual property rights that the TRIPS Agreement regulates less heavily.

INTRODUCTION

The recent enactment of the Leahy-Smith America Invents Act (AIA)[1] demonstrates congressional interest in the patent system. Most of the provisions of the AIA apply to any type of patented invention. For example, the first-inventor-to-file priority system,[2] prior user rights,[3] and post-grant and *inter partes* review proceedings[4] apply equally to chemical compounds, electrical appliances, mechanical devices, and any other invention that may be protected by a patent. However, other AIA provisions are specific to particular types of inventions. That statute limited the availability of patents on tax strategies,[5] prohibits the issuance of patents claiming human organisms,[6] and creates "transitional proceedings" that apply exclusively to patents pertaining to business methods.[7] The AIA also allows "prioritization of examination of applications for products, processes, or technologies that are important to the national economy or national competitiveness.... "[8]

The AIA reflects the principle that, for the most part, the U.S. patent system operates in a uniform manner. All patentable inventions are generally subject to the same statutory provisions.[9] A number of exceptions exist to this concept of technological neutrality, however. For example, statutory provisions limit the enforceability of patents claiming methods of medical treatment;[10] call for patent term extension for certain products regulated by the Food and Drug Administration (FDA);[11] and establish specialized patents for designs and plants.[12]

This blended architecture has for many years prompted inquiry into whether the patent system operates best as a uniform system that applies neutrally to all inventions, or whether it could or should be tailored to meet the specific needs of different industries.[13] Commentators have proposed, for example, that software patents should receive shorter terms than patents on other inventions,[14] and that patents on genes should be subject to compulsory licenses that allow individuals to use the patented technology upon paying a license fee.[15] Unenacted legislation in the 112th Congress, H.R. 6245, proposed patent litigation reforms that would apply to patents claiming computer software and hardware, but not to other sorts of inventions.

Notably, U.S. membership in the World Trade Organization (WTO) may influence congressional willingness to promulgate industry-specific patent statutes. One component of the WTO agreements, the Agreement on Trade-Related Aspects of Intellectual Property Rights (TRIPS Agreement),[16] requires patents to be available and enforceable "without discrimination as to ... the field of technology.... "[17] Although the TRIPS Agreement allows for some limited exceptions to this rule of technological neutrality,[18] one consequence of U.S. membership in the WTO may be a restricted ability to tailor the patent statute to particular inventions and industries.[19]

This report considers the possibility of modifying the U.S. patent system to meet the needs of specific industries. After providing a brief review of the patent system, the report identifies different industrial traits, such as the pace of innovation, product cycle, and the cost of research and development, which potentially suggest the desirability of tailored patent rights. It then considers possible points of adjustment within the patent system, including patent term, scope of exclusive rights, and the availability of remedies for infringement. The report then discusses potential difficulties associated with distinguishing among industries within the patent system, including the provisions of the TRIPS Agreement. The report closes with concluding observations.

PATENT FUNDAMENTALS

The patent system covers a broad variety of inventions.[20] U.S. patents cover traditional subject matter such as machines, pharmaceuticals, and manufacturing processes, along with high-tech inventions in the fields of biotechnology, computer software, and nanotechnology. No matter what the sort of technology, however, the patent system in large measure operates under the same general principles.

In particular, all inventors who wish to obtain patent rights must file an application at the U.S. Patent and Trademark Office (USPTO). USPTO examiners then review the application to ensure that certain statutory requirements are met. These statutory requirements—including that the invention be adequately described in the patent application,[21] and that the invention must not have been obvious to a skilled artisan[22]—apply to every invention for which a patent is sought. Similarly, once a patent has issued, the statutory term is set uniformly to 20 years from the date of filing no matter what the type of invention.[23] All patents received the identical exclusive rights—namely, the right to prevent others from making, using, selling, offering to sell, or importing the patented invention in the United States.[24]

Some exceptions exist to this general notion of standardization. The United States has long provided for the protection of industrial designs through so-called "design patents."[25] The Patent Law Treaties Implementation Act of 2012, P.L. 112-211, recently caused U.S. law to conform to the Hague Agreement Concerning the International Registration of Industrial Designs, the leading international agreement on these specialized rights. The United States also established two sorts of intellectual property rights for plants: plant patents, which are administered by the USPTO;[26] and plant variety protection certificates, which are administered by the Department of Agriculture.[27] In addition, the Hatch-Waxman Act extends the terms of patents covering pharmaceuticals, medical devices, and other products which are subject to premarketing approval by the FDA.[28] As well, over the years private legislation has increased the terms of a number of particular patents.[29]

Further, some commentators observe that the statutory provisions for patents are stated fairly generally. In their view, these broad parameters provide courts and the USPTO with some ability to recognize the diverse characteristics of distinct technologies and tailor their rulings accordingly.[30] For example, one core patentability requirement is that the invention must not have been obvious to a person of ordinary skill in the art.[31] The courts have arguably developed nuanced obviousness principles that take into account whether the invention falls within a predictable art, including many mechanical and electrical inventions, as compared to a less predictable art, such as some branches of chemistry and biotechnology. Inventions within these latter, less predictable arts may possibly be more readily able to satisfy the obviousness requirement and therefore are more likely to be patented.[32]

Despite these exceptions, the Patent Act of 1952, as amended by such legislation as the American Inventors Protection Act[33] and AIA, is generally worded in a neutral fashion.[34] The uniform treatment provided by the patent system may be contrasted with the diversity of the technologies and industries to which patents pertain, however, a topic this report takes up next.

THE ROLE OF PATENTS WITHIN DIFFERENT INDUSTRIES

Commentators have for many years recognized that technologies and industries vary in ways that are salient to the patent system. In particular, the costs and risks of conducting R&D differ widely among industries. Arguably the patent incentive should be greater for innovative efforts that are expensive and more likely to fail.[35] Similarly, some sectors are marked by stand-alone, discrete advances, while others feature steady, cumulative innovation. Firms that operate within the latter sorts of industries may have greater need to access technologies invented and patented by others.[36]

As well, product cycles also vary markedly across the U.S. economy. For example, patients may use a particular pharmaceutical for decades, even as consumers quickly discard their old mobile telephones and other electronics in favor of new devices. Industries with compact product cycles may be better served by short application pendency periods at the USPTO and also may be less concerned with lengthy terms of patent protection.[37]

Still another factor may be the practical availability of trade secret protection.[38] Competitors may readily reverse engineer many electrical and mechanical products, for example. Firms within these industries must therefore seek patent rights or enjoy no effective intellectual property rights whatsoever. On the other hand, certain chemical and biotechnology inventions may be more plausibly maintained as trade secrets. Arguably greater encouragement is needed for public disclosure of chemical and biological inventions through the patent system.[39]

The number of patents that cover a particular product also varies widely among industry. Within the pharmaceutical industry, often only a handful, and sometimes only one or two patents cover an individual drug. In contrast, numerous patents may cover a particular electronics product. As RPX Corporation, a self-described "comprehensive patent risk management firm," explains:

> Modern products and services incorporate numerous technology components. The evolution of mobile devices provides an example. Based on our research, we believe there are more than 250,000 active patents relevant to today's smartphones, a significant increase compared to our estimate of approximately 70,000 patents that were active and relevant to mobile phones in 2000. This growth can be attributed to the expanded set of features and functionality incorporated in today's smartphones, including touchscreens, internet access, streaming video, media playback, application store readiness and other web-based services, and WiFi connectivity options.[40]

These "vast disparities in patent-to-product" ratios may influence the perceptions of different business enterprises over the value of individual patents as compared to larger portfolios of intellectual property rights.[41]

Another arguable distinction among industries is the detectability of patents that are pertinent to particular products.[42] Fields such as small-molecule chemistry employ an internationally recognized, standardized nomenclature to identify the composition of a particular compound. As a result, firms within the agricultural, industrial, and pharmaceutical chemical industries may enjoy some confidence that they will be able to identify patents that are relevant to products they wish to sell.[43] In contrast, fields such as software arguably have yet to develop a standardized terminology for identifying programs, procedures, algorithms, and other sorts of machine instructions. The lack of conventions for identifying software inventions potentially impacts the patent system, arguably making the identification of pertinent patents difficult for firms in the computer industry.[44]

Implications for the Patent System

In view of these and possibly other distinctions between technologies and markets, policymakers may not be particularly confident that a one-size-fits-all patent system ideally encourages innovation throughout a diverse array of industries.[45] Observers have for many years suggested that the statute distinguish among the varying inventions and industries for which patents are available. In theory, many aspects of the patent system could be tailored to reflect the needs of distinct technology sectors. Some possible points of adjustment include

- The speed with which the USPTO reviews patent applications could be adjusted to meet the needs of industries with varying product cycles.
- Patents could be more difficult to obtain within some fields than others depending upon the perceived need of the patent incentive to encourage innovation.
- Patents on inventions capable of being protected by trade secrets in whole or in part—for example, biotechnologies or chemical processes—might be required to have more complete technical disclosures in order to allow others to practice the invention expeditiously.
- The term of protection could vary among technologies depending upon the pace of innovation within particular fields.
- The scope of exclusive rights could be adjusted in order to immunize select individuals or enterprises—such as physicians or universities—from liability for patent infringement.
- Compulsory licenses could be made available for categories of patented inventions perceived to fulfill an important public need—for example, healthcare or environmental technologies—thereby allowing third parties to use the invention without the permission of the patent proprietor.

Congress has implemented some of these technology-specific measures. Among other examples, the Hatch-Waxman Act extends the terms of patents covering pharmaceuticals, medical devices, and other products which are subject to premarketing approval.[46] Congress also imposed restrictions upon the enforceability of patents claiming medical procedures.[47] Specialized patent rights exist both for industrial designs and plants.[48]

More recently, the deliberations that led to the enactment of the Leahy-Smith America Invents Act (AIA)[49] again placed focus upon the role of the patent system within different industries. One topic that focused attention upon these distinctions concerned the award of remedies when firms or individuals are found to infringe patents. Broadly speaking, the patent law allows courts to award two sorts of remedies against infringers: (1) injunctions, or court orders requiring an entity to cease future infringements;[50] and (2) damages, consisting of an award of money to compensate patent proprietors for financial losses suffered due to infringements.[51] The legislative history of the AIA reveals that certain representatives of the pharmaceutical and information technology sectors often held different views about the state of the law of patent remedies.[52]

With respect to injunctions, prior to 2006 courts would virtually always enjoin an adjudicated infringer from future practice of the patented invention. Some observers—particularly those from the information technology sector—believed that this "automatic injunction" rule was particularly unfair when the patented invention supported a single function in a multifunctional product, such as an automobile, smart phone, or computer software.[53] They also believed that this rule encouraged the assertion of patents of dubious quality and offered patent proprietors too much leverage during licensing and settlement negotiations.[54] However, other observers—primarily from the pharmaceutical sector—asserted that absent the award of an injunction, competitors would not respect the exclusive rights afforded by a patent.[55]

A predecessor, unenacted version of the AIA, the Patent Reform Act of 2005,[56] proposed changes to the principles governing injunctions in patent law. Section 7 of that bill provided that in deciding whether to issue an injunction or not, "the court shall consider the fairness of the remedy in light of all the facts and the relevant interests of the parties associated with the invention." Perhaps because of the discordant view of different patent stakeholders as to the availability of injunctions in the patent community, this provision was identified as one of the most controversial within the bill.[57]

While debate about congressional action with respect to injunctions in patent law continued, the Supreme Court issued its decision in *eBay Inc. v. MercExchange, L.L.C.* in 2006.[58] There the Court unanimously held that an injunction should not automatically issue based on a finding of patent infringement. Under the *eBay* ruling, courts must weigh equitable factors traditionally used to determine if an injunction should issue, including whether the patent proprietor suffered an irreparable injury; the award of damages would be inadequate to compensate for that injury; that considering the balance of hardships between the patent owner and infringer, an injunction is warranted; and that the public interest would not be disserved by a permanent injunction.[59]

Following the *eBay* decision, Congress did not include legislative provisions concerning injunctive relief in the AIA. Some commentators believe that the Supreme Court struck an appropriate balance by addressing concerns of information technology firms that were concerned that patent holders would assert their rights after the launch of a commercially successful product in an effort to extort an unreasonably large royalty using the threat of an injunction; but also recognized that most often a prevailing patent holder would be awarded an injunction.[60] Further, the early prediction that "the biotech and pharmaceutical industries have little to fear in the post*eBay* world"[61] appears to have been correct, for the great majority of judicial decisions declining the award of a permanent injunction have concerned information technology patents.[62]

A second, arguably divisive issue between the information technology and pharmaceutical industries pertains to damages. Marketplace realities often render the determination of an appropriate damages award a difficult affair in patent litigation. In some cases, the product or process that is found to infringe may incorporate numerous additional elements beyond the patented invention. For example, the asserted patent may relate to a single component of a touch screen display, while the accused product consists of an entire smartphone. In such circumstances, a court may apply "the entire market value rule,"[63] which "permits recovery of damages based upon the entire apparatus containing several features, where the patent-related feature is the basis for consumer demand."[64] On the other hand, if the court determines that the infringing sales were due to many factors beyond the use of the patented invention, the court may apply principles of "apportionment" to reach a just measure of damages for infringement.[65]

Some believe that current damages standards have resulted in the systemic overcompensation of patent owners.[66] Such overcompensation may place unreasonable royalty burdens upon producers of high technology products, ultimately impeding the process of technological innovation and dissemination that the patent system is meant to foster. Others believe that current case law appropriately accounts for apportionment concerns. These observers are concerned that this reform might overly restrict damages in patent cases, thereby discouraging voluntary licensing and promoting infringement of patent rights. Limited damage awards for patent infringement might prevent innovators from realizing the value of their inventive contributions, a principal goal of the patent system.[67]

As discussion of damages reform has proceeded before Congress, the courts have also been active. One of the more notable cases on patent damages principles arose from the efforts of Lucent Technologies, Inc., to enforce its so-called "Day patent," which related to a method of entering information into fields on a computer screen without using a keyboard.[68] In 2002, Lucent brought an infringement suit against computer manufacturer Gateway, Inc., asserting that certain pre-installed Microsoft software infringed the Day patent. At trial, the jury found the Day patent not invalid and infringed. Lucent sought damages of $561.9 million based on 8% of Microsoft's infringing sales, while Microsoft asserted "that a lump-sum payment of $6.5 million would have been the correct amount for licensing the protected technology." The jury then awarded Lucent a single lump-sum amount of $357.7 million.[69]

The litigation in *Lucent v. Gateway* captured the attention of many observers. In a March 3, 2009, letter addressed to Senator Patrick Leahy, Senator Arlen Specter requested a delay in Senate action on the pending patent reform bill until the Federal Circuit heard oral argument in the case.[70] Observing a "symbiotic relationship between the judicial and legislative branches with regard to changes to the patent system," Senator Specter believed that "oral argument has the potential to facilitate a compromise or clarify the applicability of damages theories in various contexts."[71]

In its subsequent decision, the Federal Circuit upheld the lower court's determination that the Day patent was not invalid and infringed. In the most anticipated portion of the opinion, the appellate court also struck down the jury's damages award as not supported by substantial evidence.[72]

Some commentators have viewed the lengthy opinion as doing much to dampen speculative damages awards, particularly with respect to patents from the information technology sector.[73] Following that opinion, Lucent and Microsoft have reportedly settled the litigation.[74] Of broader interest, Congress also chose not to address damages for patent infringement within the AIA.

The timely issuance of *eBay* and *Lucent v. Gateway* may have contributed to congressional belief that courts were appropriately addressing injunctions and damages, respectively. As a result, Congress did not endeavor to reconcile the perceived needs of industries with different innovation and marketplace environments. Another legislative possibility, at least in theory, was to enact distinct remedial rules for different categories of inventions. This report next explores whether industry-specific rules are a practical possibility for patent legislation.

THE FEASIBILITY OF SECTOR-SPECIFIC PATENT RULES

In view of the concerns noted above, commentators have gone so far to say that "it has become increasingly difficult to believe that a one-size-fits-all approach to patent law can survive."[75] To the extent the current patent system creates a blanket set of rules that apply comparably to distinct industries, it likely over-encourages innovation in some contexts and under-incentivizes it in others.[76] Further, some observers have asserted that the need of firms to identify and access the patented inventions of others may differ among industries.[77] As a result, the case can be made that distinct industrial, technological, and market characteristics that exist across the breadth of the U.S. economy compel industry-specific patent statutes.

However, others have questioned the wisdom and practicality of such line-drawing.[78] The following concerns, among others, have been identified:

- Over its long history, the U.S. patent system has flexibly adapted to new technologies such as biotechnology and computer software. Legislative adoption of technology-specific categories may leave unanticipated, cutting-edge technologies outside the patent system.[79]
- Defining a specific industry or category of technologies may prove to be a contested proposition.[80]
- Over time, new industries may emerge and old industries may consolidate. The dynamic nature of the U.S. economy suggests greater need for legislative oversight within a differentiated patent regime.[81]
- Even if an industry or technology remains relatively stable, the innovation environment within it might change. For example, technological or scientific advances might open new possibilities for research and development within hidebound industries—but also increase expense and risk for those firms.[82]
- Distinct patent rights among industries or technologies may lead to strategic behavior on behalf of patent applicants. For example, a computer program that controls a fuel injector within an automobile could possibly be identified as either an automobile-related or a computer-related invention.[83]

- The legislative effort to enact sector-specific patent laws may provide an opportunity for politically savvy firms to exert more lobbying and political power, at the possible expense of less sophisticated firms.[84]

In addition to these practical concerns, U.S. membership in the World Trade Organization (WTO) may restrict congressional ability to tailor the patent system to account for different industries and inventions, to the extent that compliance with WTO standards is desired. The Agreement on Trade-Related Aspects of Intellectual Property Rights (TRIPS Agreement) is a WTO-administered treaty that stipulates minimum standards for many forms of intellectual property protection.[85] Article 27, paragraph one of the TRIPS Agreement expressly provides that, with some exceptions, "patents shall be available and patent rights enjoyable without discrimination as to the place of invention, the field of technology and whether products are imported or locally produced."

At the time it was drafted, the TRIPS Agreement standard of technological neutrality was arguably intended to provide for patent protection for a variety of inventions for which patents were previously unavailable in many countries.[86] For example, certain jurisdictions did not allow patents to issue on pharmaceuticals and agricultural chemicals prior to joining the WTO. The wording of Article 27.1 appears to have broader implications than merely requiring WTO member states to grant patents on pharmaceuticals and other previously unpatentable inventions, however. Its principle of non-discrimination seems broadly to require that all inventions in all fields of technology be treated identically—a reading that would block, for example, discriminating in favor of particular inventions as well as against them.[87]

Article 27 expressly permits some limited exceptions to this concept of homogeneity. In particular, WTO members may exclude from patentability inventions whose commercial exploitation would violate the public order or morality. However, inventions may not be exempted from patentability merely because their use is illegal.[88] In addition, diagnostic, therapeutic, and surgical methods may also be excluded from patentability.[89] In addition, WTO member states may deny patents on "plants and animals other than micro-organisms, and essentially biological processes for the production of plants or animals other than non-biological and microbiological processes." However, if a WTO member state opts to do so, it must protect plant varieties through an alternative specialized system, typically known as "plant breeder's rights."[90] In addition, the TRIPS Agreement allows for "security exceptions" that permit

WTO members to take any action considered necessary for the protection of "essential security interests."[91]

In addition, Article 30 of the TRIPS Agreement allows for "exceptions to rights conferred." The TRIPS Agreement stipulates that each WTO member state must provide patent proprietors with the right to exclude others from making, using, selling, offering to sell, and importing the patented invention.[92] Article 30 then reads:

> Members may provide limited exceptions to the exclusive rights conferred by a patent, provided that such exceptions do not unreasonably conflict with a normal exploitation of the patent and do not unreasonably prejudice the legitimate interests of the patent owner, taking account of the legitimate interests of third parties.

Article 30 appears to contemplate limited exceptions to the patent rights for such reasons as scientific experiments or use of a patented product for purposes of obtaining regulatory approval.[93]

At first glance, Article 30 of the TRIPS Agreement might seem to provide WTO member states with the ability to provide technology-specific exceptions despite the wording of Article 27.1. However, one dispute settlement panel of the WTO has "concluded ... that the anti-discrimination rule of Article 27.1 does apply to exceptions of the kind authorized by Article 30."[94] Stated differently, the WTO has reasoned that even exceptions to a patent owner's exclusive rights must be technology neutral.

The WTO's conclusion that even exceptions to a patent holder's exclusive rights must operate in a uniform manner has attracted some critical commentary. On one hand, if an exception to a patent holder's exclusive rights must apply to all technologies and industries, it is less likely to be "limited" and therefore compatible with Article 27.1.[95] On the other hand, the WTO reasoned that "the TRIPS Agreement [required] governments to apply exceptions in a nondiscriminatory manner, in order to ensure that governments do not succumb to domestic pressures to limit exceptions to areas where right holders tend to be foreign producers."[96]

The WTO dispute resolution panel in the *Canada Pharmaceutical* case further observed:

> [I]t is not true that Article 27 requires all Article 30 exceptions to be applied to all products. Article 27 prohibits only discrimination as to the place of invention, the field of technology, and whether products are

imported or produced locally. Article 27 does not prohibit bona fide exceptions to deal with problems that may exist only in certain product areas.[97]

The distinction drawn by the WTO panel between the terms "field of technology" and "certain product areas" arguably does not shimmer with clarity. The panel's opinion obtains no further guidance as to the meaning of "certain product areas." However, some commentators have suggested that patent statutes may permissibly draw distinctions based on narrow categories of goods or services, in contrast to broader fields of technological endeavor. For example, Wesley A. Cann, Jr., an emeritus member of the University of Connecticut business school faculty, has asserted that "[e]ven if Article 30 would not allow a limited exception to be directed at the entire pharmaceutical industry (a position that is still open to substantial doubt), it can be argued that an exception could be made for those particular pharmaceuticals aimed at the prevention and treatment of HIV/AIDS."[98]

As a result, the membership of the United States within the WTO provides a possible constraint against tailoring the patent system to meet the perceived needs of specific industries. To the extent that WTO compliance is desired, the U.S. patent system must ordinarily act in a homogenous manner with respect to particular fields of technology. The TRIPS Agreement does permit some specific exceptions, however, for such subject matter as diagnostic, therapeutic, and surgical methods; matters of national security; and, as explained by the *Canada Pharmaceutical* WTO panel, "certain product areas."

CONCLUDING OBSERVATIONS

The patent system involves numerous parameters, including the requirements to obtain a patent, the scope of proprietary rights, and the term of protection. In theory these attributes could be tailored to meet the needs of different sectors. Defining industry-specific or technology-specific patent doctrines presents some practical difficulties, however, and may also give rise to incompatibilities with the WTO TRIPS Agreement.

Should Congress consider current circumstances with respect to the patent system to be appropriate, then no action need be taken. However, should Congress believe that different sectors possess distinctive needs with respect to the patent system, a number of options exist. As noted, the TRIPS Agreement provides for a number of limited exceptions to the general rule of technological neutrality. U.S. legislation could simply track the exceptions identified by the provisions of the TRIPS Agreement, with particular reference to their interpretation by WTO dispute resolution panels.

Congress could also potentially draw distinctions among grounds not identified by Article 27.1 of the TRIPS Agreement—namely, the place of invention, the field of technology, and whether products are imported or produced locally. For example, the Leahy-Smith America Invents Act (AIA) established an infringement defense based upon the prior commercial use of the patented invention by the accused infringer.[99] The AIA stipulated that this defense was unavailable if the patented invention was made by an institution of higher education, however.[100] This "university exception" appears to comply with the TRIPS Agreement because it draws a distinction based upon the identity of the inventor, rather than on a ground identified in Article 27.1.

Another option is the establishment of additional intellectual property rights that complement patents. For example, Congress has established so-called "regulatory exclusivities," a term that refers to a period of time during which a regulated product is afforded protection from competing applications for marketing approval.[101] Most notably, the food and drug laws establish a number of regulatory exclusivities that apply to pharmaceuticals and are administered by the Food and Drug Administration. In contrast to patents, regulatory exclusivities are not subject to extensive regulation by the TRIPS Agreement.[102] As a result, at least with respect to regulated industries, regulatory exclusivities may provide a more flexible instrument for legislative tailoring of intellectual property rights.[103]

Whether legal doctrines should be expressed as discrete rules or more generally phrased standards is a long-standing debate within the field of jurisprudence.[104] Placed within the context of congressional consideration of our intellectual property laws, the debate suggests that distinct rules for different sectors form one possibility for patent reform. Yet Congress may also craft broader principles that may be appropriately applied across the diverse industries and technologies to which the patent system pertains. Determining whether a rule or a standard provides the most appropriate vehicle for a particular reform to our patent laws remains a matter of legislative judgment.

End Notes

[1] P.L. 112-29, 125 Stat. 284 (2011). See CRS Report R42014, The Leahy-Smith America Invents Act: Innovation Issues, by John R. Thomas.
[2] Ibid. at §3.
[3] Ibid. at §5.
[4] Ibid. at §6.
[5] Ibid. at §14.
[6] Ibid. at §33.
[7] Ibid. at §18.
[8] Ibid. at §25.
[9] Clarisa Long, "Our Uniform Patent System," 55 The Federal Lawyer (2008), 44.
[10] 35 U.S.C. §287(c).
[11] 35 U.S.C. §156.
[12] 35 U.S.C. §§161, 171.
[13] Michael W. Carroll, "One Size Does Not Fit All: A Framework for Tailoring Intellectual Property Rights," 70 Ohio State Law Journal (2009), 1361.
[14] Mark H. Webbink, "A New Paradigm for Intellectual Property Rights in Software," Duke Law & Technology Review (2005), 12.
[15] Daniel M. Lorentzen, "Do These Genes Fit? Genes as Patentable Subject Matter," 60 Drake Law Review (2012), 933.
[16] Agreement on Trade-Related Aspects of Intellectual Property Rights, April 15, 1994, Marrakesh Agreement Establishing the World Trade Organization, Annex 1C, 1869 U.N.T.S. 299 (1994).
[17] Ibid. at Art. 27.
[18] Ibid.
[19] Dan L. Burk & Mark A. Lemley, "Is Patent Law Technology-Specific?," 17 Berkeley Technology Law Journal (2002) 1155.
[20] 35 U.S.C. §101 (patents available on machines, manufactures, compositions of matter, and processes).
[21] 35 U.S.C. §112(a).
[22] 35 U.S.C. §103.
[23] 35 U.S.C. §154(a)(2).
[24] 35 U.S.C. §271(a).
[25] 35 U.S.C. §171-173.
[26] 35 U.S.C. §§161-164.
[27] 7 U.S.C. §§2321-2582.
[28] 35 U.S.C. §156. See CRS Report R41114, The Hatch-Waxman Act: Over a Quarter Century Later, by Wendy H. Schacht and John R. Thomas.
[29] See CRS Report RS21129, Pharmaceutical Patent Term Extensions: A Brief Explanation, by Wendy H. Schacht and John R. Thomas.
[30] Dan L. Burk and Mark A. Lemley, "Policy Levers in Patent Law," 89 Virginia Law Review (2003), 1575.
[31] 35 U.S.C. §103(a).
[32] See Briana Barron, "Structural Uncertainty: Understanding the Federal Circuit's Lead Compound Analysis," 16 Marquette Intellectual Property Law Review (Summer 2012), 401.
[33] P.L. 106-113.

[34] Michael W. Carroll, "One for All: The Problem of Uniformity Cost in Intellectual Property Law," 55 American University Law Review (2006), 845.

[35] For discussion of how the features of a patent system might pertain to R&D costs, see Robert M. Hunt, "Economics and the Design of Patent Systems," 13 Michigan Telecommunications & Technology Law Review (Spring 2007), 457.

[36] See Mark A. Lemley, "The Myth of the Sole Inventor," 110 Michigan Law Review (2012), 709 (discussing how cumulative innovation witnessed in many industries may impact patent law principles).

[37] See Sarah A. Geers, "Common Sense and the Fact Finder Without Skill in the Art: The Role of Objective Evidence in Achieving Proper Technology Specificity," 40 Seton Hall Law Review (2010), 425 (discussing patent system design in relation to distinct industry product cycles).

[38] See CRS Report R41391, The Role of Trade Secrets in Innovation Policy, by John R. Thomas.

[39] For a general discussion of the relationship between patents and trade secrets, see J. Jonas Anderson, "Secret Inventions," 26 Berkeley Technology Law Journal (2011), 917.

[40] See Daniel O'Connor, "One in Six Active U.S. Patents Pertain to the Smartphone," Disruptive Competition Project (October 17, 2012) (available at http://www.project-disco.org/intellectual-property/one-in-six-active-u-s-patentspertain-to-the-smartphone/).

[41] Graeme B. Dinwoodie and Rochelle C. Dreyfuss, "Diversifying Without Discriminating: Complying with the Mandates of the TRIPS Agreement," 13 Michigan Telecommunications & Technology Law Review (2007), 445.

[42] See James Bessen and Michael J. Meurer, Patent Failure: How Judges, Bureaucrats, and Lawyers Put Innovation at Risk (Princeton University Press 2008), 46-72.

[43] See Christa J. Laser, "A Definite Claim on Claim Indefiniteness: An Empirical Study of Definiteness Cases of the Past Decade with a Focus on the Federal Circuit and the Insolubly Ambiguous Standard," 10 Chicago-Kent Journal of Intellectual Property (2010), 25.

[44] Ibid.

[45] Dan L. Burk and Mark A. Lemley, "Is Patent Law Technology-Specific?," 17 Berkeley Technology Law Journal (2002) 1155.

[46] 35 U.S.C. §156.

[47] 35 U.S.C. §287(c).

[48] 35 U.S.C. §§171-173 (designs); 35 U.S.C. §§161-164 (plants).

[49] P.L. 112-29, 125 Stat. 284 (2011).

[50] 35 U.S.C. §283.

[51] 35 U.S.C. §284.

[52] See CRS Report RL33367, Patent Reform: Issues in the Biomedical and Software Industries, by Wendy H. Schacht.

[53] See Lea Shaver, "Illuminating Innovation: From Patent Racing to Patent War," 69 Washington & Lee Law Review (2012), 1891.

[54] Nicholas P. Chan, "Balancing Judicial Misvaluation and Patent Hold-Up: Some Principles for Considering Injunctive Relief After Ebay," 59 UCLA Law Review (2012), 746.

[55] Paul M. Schoenhard, "Who Took My IP? Defending the Availability of Injunctive Relief for Patent Owners," 16 Texas Intellectual Property Law Journal (2008), 187.

[56] H.R. 2795, 109th Congress.

[57] George H. Pike, "The Next Step in Patent Reform," 28 Information Today no. 10 (November 1, 2011), 1.

[58] 547 U.S. 388 (2006).

[59] See Mark P. Gergen et al., "The Supreme Court's Accidental Revolution: The Test for Permanent Injunctions," 112 Columbia Law Review (2012), 203.

[60] William C. Rooklidge and Alyson G. Barker, "Reform of a Fast-Moving Target: The Development of Patent Law Since the 2004 National Academies Report," 91 Journal of the Patent and Trademark Office Society (2009), 153.

[61] Michael Beylkin, "Much Ado About Nothing: The Biotech and Pharmaceutical Industries Have Little to Fear in the Post-eBay World," 6 Journal of Telecommunications & High Technology Law (2007), 179.

[62] See University of Houston Law Center, PatStats: U.S Patent Litigation Statistics (available at http://www.patstats.org).

[63] LaserDynamics, Inc. v. Quanta Computer, Inc., 694 F.3d 51 (Fed. Cir. 2012).

[64] State Industries, Inc. v. Mor-Flo Industries, Inc., 883 F.2d 1573 (Fed. Cir. 1989).

[65] See LaserDynamics v. Quanta, supra.

[66] See Amy L. Landers, "Let the Games Begin: Incentives to Innovation in the New Economy of Intellectual Property Law," 46 Santa Clara Law Review (2006), 307.

[67] See William C. Rooklidge, "Reform of the Patent Laws: Forging Legislation Addressing Disparate Interests," 88 Journal of the Patent and Trademark Office Society (2006), 9.

[68] Lucent Technologies, Inc. v. Gateway, Inc., 580 F.3d 1301 (Fed. Cir. 2009).

[69] Ibid. at 1308.

[70] The letter is available at http://www.fr.com/files/uploads/attachments//patentdamages//03-03-09-Specter-to-PJLPatents.pdf.

[71] Ibid.

[72] 580 F.3d at 1335.

[73] See Bo Zeng, "Lucent v. Gateway: Putting the "Reasonable" Back into Reasonable Royalties," 26 Berkeley Technology Law Journal (2011), 329.

[74] "Microsoft and Alcatel-Lucent Settle Patent Dispute," Computer News Middle East (January 19, 2012).

[75] Dreyfuss and Dinwoodie, supra.

[76] See Alan Devlin, "Patent Law's Parsimony Principle," 25 Berkeley Technology Law Journal (2010), 1693.

[77] See Alan Devlin, "Systematic Bias in Patent Law," 61 DePaul Law Review (2011), 57.

[78] See Luigi Orsenigo and Valerio Sterzi, Comparative Study of the Use of Patents in Different Industries, Knowledge, Internationalization and Technologies Studies Working Paper 33/2010 (2010).

[79] Ibid.

[80] See Lewis Kaplow, "Why (Ever) Define Markets?," 124 Harvard Law Review (2010), 437.

[81] Burk and Lemley, supra.

[82] Orsenigo and Sterzi, supra.

[83] See Kristen Osenga, "The Shape of Things to Come: What We Can Learn About Patent Claim Length," 28 Santa Clara Computer & High Technology Law Journal (2012), 617.

[84] Burk and Lemley, supra.

[85] Agreement on Trade-Related Aspects of Intellectual Property Rights, April 15, 1994, Marrakesh Agreement Establishing the World Trade Organization, Annex 1C, 1869 U.N.T.S. 299, 33 I.L.M. 1197 (1994).

[86] See Robert Weissman, "A Long, Strange Trips: The Pharmaceutical Industry Drive to Harmonize Global Intellectual Property Rules, and the Remaining WTO Legal Alternatives Available to Third World Countries," 17 University of Pennsylvania Journal of International Economic Law (1996), 1069.

[87] See Rebecca S. Eisenberg, "The Role of the FDA in Innovation Policy," 13 Michigan Telecommunications & Technology Law Review (2007), 345.

[88] TRIPS Art. 27.2.

[89] Ibid. at Art. 27.3(a).

[90] Ibid. at Art. 27.3(b).

[91] Ibid. at Art. 73.

[92] Ibid. at Art. 28.

[93] See Scott Taylor, "Where Are the Green Machines? Using the Patent System to Encourage Green Invention and Technology Transfer," 23 Georgetown International Environmental Law Review (2011), 577.

[94] Panel Report, Canada-Patent Protection of Pharmaceutical Products, WT/DS114/R (March 17, 2000) (hereinafter "Canada Pharmaceutical").

[95] See Dinwoodie and Dreyfuss, supra.

[96] Canada Pharmaceutical, paragraph 7.92.

[97] Ibid.

[98] Wesley A. Cann, Jr., "On the Relationship Between Intellectual Property Rights and the Need of Less-Developed Countries for Access to Pharmaceuticals: Creating a Legal Duty to Supply Under a Theory of Progressive Global Constitutionalism," 25 University of Pennsylvania Journal of International Economic Law (2004), 755.

[99] P.L. 112-29, 125 Stat. 284 (2011), at §5.

[100] Ibid. (establishing 35 U.S.C. §273(e)(5)).

[101] See Yaniv Heled, "Patents vs. Statutory Exclusivities in Biological Pharmaceuticals—Do We Really Need Both?," 18 Michigan Telecommunications & Technology Law Review (2012), 419.

[102] See Christina Bucci, "Responsible Patent Protections: Preserving Public Health Objectives in the Trans-Pacific Partnership Agreement," 26 Pacific McGeorge Global Business & Development Law Journal (2013), 213.

[103] See Eisenberg, supra.

[104] See David Olson & Stefania Fusco, "Rules Versus Standards: Competing Notions of Inconsistency Robustness in Patent Law," 64 Alabama Law Review (2013), 647.

INDEX

C

F

G

H

I

J

L

M

N

O

P

Q

T

U

V

W

Y